COMMUNITY LIBRARY
3 2301 00167395 7

D0820311

The Lives of Ants

The Lives of ANTS

by Laurent Keller and Élisabeth Gordon

translated by James Grieve

OXFORD
UNIVERSITY PRESS

OXFORD
UNIVERSITY PRESS

Great Clarendon Street, Oxford OX2 6DP

Oxford University Press is a department of the University of Oxford.
It furthers the University's objective of excellence in research, scholarship,
and education by publishing worldwide in

Oxford New York

Auckland Cape Town Dar es Salaam Hong Kong Karachi
Kuala Lumpur Madrid Melbourne Mexico City Nairobi
New Delhi Shanghai Taipei Toronto

With offices in

Argentina Austria Brazil Chile Czech Republic France Greece
Guatemala Hungary Italy Japan Poland Portugal Singapore
South Korea Switzerland Thailand Turkey Ukraine Vietnam

Oxford is a registered trade mark of Oxford University Press
in the UK and in certain other countries

Published in the United States
by Oxford University Press Inc., New York

First published 2009

British Library Cataloguing in Publication Data
Data available

Library of Congress Cataloging in Publication Data
Library of Congress Control Number: 2008943416

Typeset by SPI Publisher Services, Pondicherry, India
Printed in Great Britain
on acid-free paper by
CPI Antony Rowe, Chippenham, Whltshire

ISBN 978–0–19–954186–7

1 3 5 7 9 10 8 6 4 2

Acknowledgements

The authors are extremely grateful to the Fondation Ernest Dubois for its generous support. Laurent Keller also acknowledges the Swiss National Foundation for its continuing support of his research. We thank Daniel Cherix, Christian Peeters, Francesco Mondada, and Gérard Jorland for their useful comments on the French version of the text. We would also like to express our gratitude to Latha Menon of Oxford University Press for her strong support and expert guidance, and to James Grieve for the difficult translation of the book. Ulrich Mueller made very valuable comments on the leaf-cutting ant section; and we are particularly grateful to Andrew Bourke for his close reading of the whole translation and his perceptive comments on it. Thanks, too, to Pierre and Chloé for having gone for so many months without their favourite veal blanquette and cheese soufflé, and to Marius and Léonore for having uncomplainingly done without all that skiing.

Contents

List of illustrations

Figure acknowledgements

Photographs in the plate section were taken by:

1. a. Alex Wild; b. Alex Wild
2. Alex Wild
3. a. Alex Wild; b. Alex Wild; c. Alex Wild; d. Alex Wild
4. a. Laurent Keller; b. Alex Wild; c. Laurent Keller; d. Alex Wild
5. a. Alex Wild; b. Alex Wild; c. Alex Wild; d. Alex Wild
6. a. Laurent Keller; b. Alex Wild; c. Daniel Kronauer; d. Daniel Kronauer
7. a. Daniel Kronauer; b. Alex Wild; c. Daniel Kronauer; d. Daniel Kronauer
8. a. Alex Wild; b. Alex Wild; c. Alex Wild
9. Alex Wild
10. Daniel Kronauer
11. a. Alex Wild; b. Alex Wild; c. Alex Wild; d. Alex Wild
12. Alex Wild
13. a. Alex Wild; b. Daniel Cherix
14. a. Christian König; b. Alex Wild; c. Elva Robinson; d. Christian König
15. a. Christian König; b. Christian König; c. Christian König
16. a. Francesco Mondada; b. Jean-Bernard Billeter & Michael Krieger; c. Francesco Mondada; d. Markus Waibel

Introduction

Since time immemorial, human beings have been fascinated, amazed, intrigued, and captivated by ants. And yet, at first glance, there is nothing particularly attractive about the tiny creatures. Unlike butterflies, they don't have wings with vivid colour patterns; they cannot boast the iridescent wing-cases seen on many beetles. Nor do they produce things which human beings like to eat or wear, such as honey or silk. They don't even chirp or sing like crickets or cicadas; and, unlike bees, they never go in for dancing.

They do, however, have other characteristics which, in their way, are much more remarkable. For one thing, their social arrangements are quite extraordinary, almost unique among living creatures, and have often been compared to human society. William Morton Wheeler, the founder of American myrmecology, wrote in *Ants* (1910): 'The resemblances between men and ants are so very conspicuous that they were noted even by aboriginal thinkers.' For another thing, ants are not only efficient, they are hard-working and thrifty, qualities which have always seemed like good reasons for seeing them as virtuous role models.

In *c.*1000 BC, King Solomon recommended them, in the Old Testament, as models of wisdom: 'Go to the ant, thou sluggard; consider her ways, and be wise: Which having no guide, overseer, or ruler, Provideth her meat in the summer, and gathereth her food in the harvest' (*Proverbs* 6:6–8). The same way of seeing them turned up centuries later in La Fontaine's fable 'The Cicada and the Ant'. They are also mentioned in the Koran, which presents them as a highly developed race of beings, and in the Talmud, again as synonymous with honesty and virtue.

The Greeks, too, Aristotle, Plato, and Plutarch, for instance, praised these social insects as wise and clever. The Roman naturalist Pliny the Elder devoted a whole chapter of his *Historia naturalis* to them, expatiating on their bravery and strength. He even mentions ants as big as dogs found in India or Ethiopia: they acted as guards outside gold-mines and killed any men who attempted to make off with the precious metal. These accounts are of course closer to fiction than to fact; but they do attest to the human appeal of ants, as well as to the fears they could engender. These figments of Antiquity's imagination show that there was an awareness of how aggressive the insects could be. But what was uppermost in the ancient world's appreciation of ants was how they could communicate with one another, devise their division of labour, and construct nests of such architectural complexity—which the natural historian Aelian compared to palatial residences.

The effect of these tiny creatures on human imagination was such that they inspired many a myth and became incorporated into belief systems. The Dogon peoples of West Africa saw them as the wives of the god Amma and the mothers of the first humans. They were also central to traditional rituals, for example among the Wayana-Apalai peoples of Brazil, Surinam, and French Guyana, where a boy reaching puberty had to demonstrate that he was worthy of adult status by wearing a sling full of fire ants round his torso or tied to his back, thus proving he had the courage and endurance to withstand the bites from these very aggressive creatures.

Literature and film

Nowadays ants have lost their previous importance in legend and ritual, but instead they figure prominently in books and films. The French writer Bernard Werber, for instance, is widely known for his best-selling *Ants* trilogy (*The Empire of the Ants*, *The Day of the Ants*, and *The Revolution of the Ants*). Ants now figure in a broad range of popular culture, from many works of science fiction, to novels, children's books, comics, and video games.

The creatures are also to be found swarming across television and cinema screens, with lead roles in many documentaries or starring in feature films. Sometimes they are presented as a threat and act out horrific fantasies, as in Gordon Douglas's *Them*, a Hollywood film of the 1950s in which mutant ants more than two metres tall spread panic throughout the United States. Usually, however, they are humanized, endowed with anthropomorphic physiques and behaviours, and as such are presented as more congenial to humankind. This is how they appear in *Antz*, directed by Eric Darnell and Tim Johnson, and *A Bug's Life* by John Lasseter. Both these animated films, which appeared in 1998, hold up a mirror to our society by having as their central characters human-like ants who feel out of place in a community of conformists, where individuality is undervalued.

The imaginations of old and young alike have been stimulated by such fictional insect worlds; and some have even developed a liking for the real thing. Toy manufacturers have taken advantage of the vogue for ants and are selling ant aquariums, so to speak, in which water is replaced by a nutritious gel, thus enabling children to have an ant colony of their own. Household pets were once goldfish and hamsters; now it's the turn of ants.

Naturalists or myrmecologists

Ants are a trendy thing nowadays; and even the scientific community is affected by the vogue. Its beginnings go back to the eighteenth century, with naturalists such as the Frenchman René Antoine Ferchault de Réaumur and the Englishman William Gould. Following in the footsteps of these illustrious predecessors, the entomologists of our own day, first and foremost the myrmecologists (the name now given to ant specialists), are enthusiastic investigators of the lives and ways of these social insects. Their research leads them into a world that is rich and full of surprises, and one that, even after decades of observation, is still full of unsolved mysteries. What is the secret of the huge ecological success of ants? How did their sociality develop? What is their social organization like in different species? The universe of the ant, when subjected to the most advanced methods of scientific investigation and observed with the magnification afforded by genetics and molecular biology, can be seen in a new light and now reveals a range of ways of life that for many years went unrecognized.

It has long been known that ants were inclined to live in intricately organized societies made up of individuals that cooperated, communicated, and divided up daily tasks. But now we also know that they have impressive abilities in finding their way and quite amazing ingenuity when it comes to building their nests, finding supplies, or exploiting other members of the animal kingdom. We can see, too, that they are capable of aggression and violence, which can disturb the apparent peace of their colonies and embroil them in fratricidal or matricidal strife. Even their sexual arrangements can be studied, which has revealed their at times strange ways of reproducing themselves and has shown the remarkable stratagems they employ so as to increase the numbers of copies of genes transferred to their descendants. In this area, as in many another, they display a marked originality. They never cease to amaze those who study them. So, welcome to the wonderful world of ants!

Part I
An Ecological Success Story

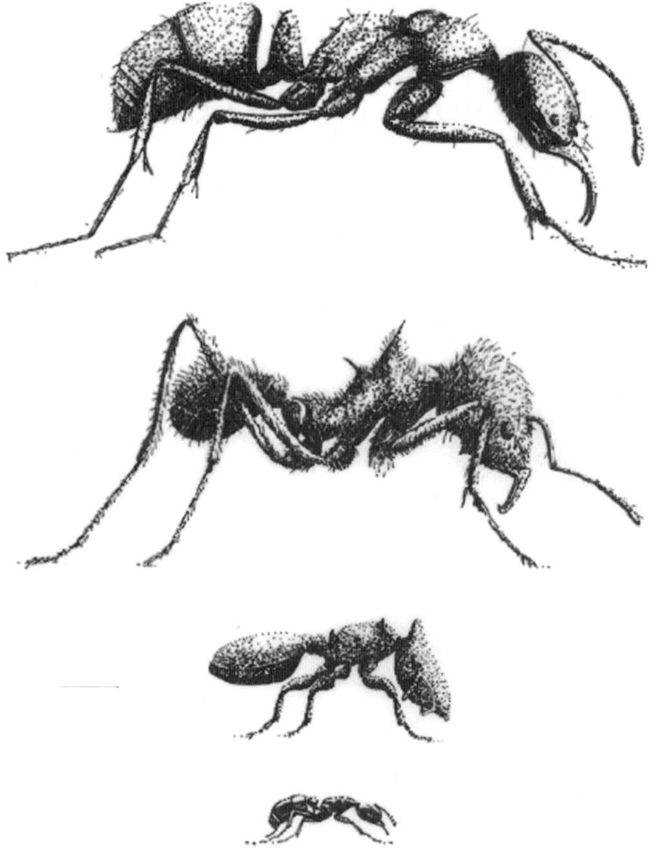

Drawing 1 Great diversity There are wide variations not only in ants' ways of life but also in their size and morphology.

From top to bottom: *Pachycondyla* (Ponerine ant), *Atta* (leaf-cutting ant), *Zacryptocerus*, *Amblyopone*.

1

Anywhere and everywhere

Ants are nothing if not ubiquitous. Venture a few steps into any woodland and there you will find a nest with all its denizens busying about. In springtime and more especially in summer, some crumbs of food or grains of sugar left lying on a draining-board will soon be visited by a long procession of tiny black creatures, scurrying along one after the other, intent on making off with an unexpected hoard of nourishment. Ants, in all their minuteness and their teeming numbers, are just part of the background, familiar, natural, taken for granted.

Yet as soon as you start to look more closely, ants turn out to be exceptional in plenty of ways. Their social organization and their ability to adapt to different environments enabled them to colonize the entire Earth some tens of millions of years ago. They greatly outnumber all other animals on the planet, including human beings.

Ants have overrun the surface of the planet. 'Ants are to be found everywhere,' says Wheeler, 'from the Arctic regions to the tropics, from the timberline on the loftiest mountains to the shifting sands of the dunes and seashores, and from the dampest forests to the driest deserts.' They are absent from only a few places, kept away by the climate. For though they can adapt to any habitat, they cannot tolerate great cold. So there are no indigenous species to be found in Greenland, Iceland, or the Antarctic. They are also notable by their absence from the forests of the North and from high altitudes (above 2,500 metres) on the wooded slopes of tropical mountains. Yet this did not prevent the great Swiss entomologist Auguste Forel (1848–1931) from leaving a description of some ants that lived at a height of 3,600 metres, collected for him by friends on the slopes of the Himalayas. Ants can even withstand great heat: out in the deserts of Africa, *Cataglyphis bicolor,* perched on its long legs, copes with temperatures close to 55 degrees.

However, their favourite habitats are the tropical regions with their abundance of fauna. In a square kilometre of rain forest, for example, there are more species of ants than there are species of primates in the whole world. Latin America boasts a very great range of ant species, as do sub-Saharan Africa and Asia. The American entomologist Edward O. Wilson and his German co-author Bert Hölldobler say in *Journey to the Ants* that, in an area of eight hectares of virgin forest in Peru, they identified more than 300 different species of ants. This number included 'forty-three species from a single tree, almost as many as occur in all of Finland, or all of the British Isles'.

This comparison with Finland is no accident, as Hölldobler has spent a long time there on the track of his favourite insects. This has led him and Wilson to the conclusion that, though ants prefer tropical climates, they manage to cope quite well in countries with temperate or even cold climates. Ants, they say, 'are the premier predators, scavengers and turners of soil in the forests of Finland', adding that they 'seldom found a patch of more than a few square metres anywhere free of these insects'. They thrive, too, in the high country of the Jura in eastern France, as well as in the Alps. Wood ants survive the winter cold by digging in a metre under the ground. In addition, they have developed another trick: they secrete glycerol and similar substances which function like antifreeze.

Millions of billions

All this means that ants are omnipresent, and more so than any other animal known to science. We do not know how many of them there are on the planet, as nobody has ever had the mind-boggling idea of totting up their numbers.

In particular places, however, scientists have been conscientious and patient enough to count the number of ants in a defined area, thus establishing that densities of ant populations can be astronomical. For instance, in a single hectare of the Ivory Coast savannah, the French entomologist Jean Levieux identified 7,000 different colonies containing in all twenty million individuals. But even this record was beaten by a group of Japanese scientists, who located a supercolony on the island of Hokkaido, composed of 306 million workers and more than a million queens, all living in a territory of 2.7 square kilometres.

On a larger scale, it would be utterly impossible to carry out a planetary census, which is why we have to fall back on an estimate of a world population of ten million billions. This figure, though only a rough approximation, is nonetheless staggering; and there can be no doubt that the population of ants far outnumbers that of humans.

Ants don't just outnumber us, they come close to outweighing us too. Tiny as they are, ten million billions of them, at an average weight of three milligrams per ant, add up to a fair weight. It actually amounts to about 10 per cent of the animal biomass, that is one-tenth of the total weight of all animals on the face of the Earth, though the percentage is higher in tropical forests. Their total weight would be roughly the same as that of the whole human population.

Family trees

Like bees and wasps, ants belong to the order Hymenoptera, within which they form a separate family, the Formicidae. Entomologists, like their zoological colleagues, like to draw up genealogical trees, so as to classify animals by similarities; and so they have divided the Formicidae first into sub-families (about twenty of them), then into genera (anywhere between 296 and 358 of these), and lastly into species. At last count, in November 2008, some 12,467 species of ants had been described, about fifty of them living in Britain, 772 in the United States, and 1,350 in Australia. Actually, these figures do not mean much; they will probably increase very soon, given the rapidity with which the myrmecologists of the world are adding to discoveries already made. Some of them have no hesitation in saying that there are many more species still unknown to science, perhaps 30,000, possibly as many as 90,000.

So, those who specialize in this field have their work cut out for them. It will consist of locating, naming, and classifying the yet undiscovered species, many of which will probably be found in the tropics, where many regions still have much to divulge about the richness of their ant life. Entomologists are convinced that, even in more temperate regions of the world and despite their own ant-like endeavours, several species of ants have still escaped detection.

2
On tastes and colours

Whether or not they belong to this or that sub-family or genus as defined by the prevailing taxonomical criteria, whether they are sub-divided into 12,000 species or many more, one thing is certain: as soon as you happen upon ants (for once you notice one of them you instantly see two or more), you recognize them straightaway. Their appearance is unmistakable: invariably the same six legs, the oblong body made of its unvarying and three clearly separate parts, the head, the thorax, and the abdomen. You could be forgiven for thinking that, if you've seen one ant, you've seen them all.

Yet, observing them through a magnifying glass, one can only but be struck by their morphological diversity. There is considerable variation between species as to how high they stand on their legs, the length of the body, and the size of the antennae. In a smallish number of species there are also certain peculiarities, such as the spikes along the body of *Polyrhachis* or the South American *Daceton*; the long-necked *Dolichoderus*, which hails from the same part of the world; or the markedly flat body of the turtle ant belonging to the genus *Zacryptocerus*. There are many other such variations on the basic theme. If you observe their heads through an electron microscope, they too show great differences in shape and mass. Seen at such close range, some of them appear quite monsterish.

There is also considerable variation in size. Take the largest known ant, the giant forest ant, *Camponotus gigas*, from Borneo: it is about three centimetres in length and its head could house an entire colony of *Brachymyrmex*, a minute ant from South Africa. What is even more surprising is that, within a single species, there can be very great differences of size, the record in this respect going to *Pheidologeton diversus* of Asia: the head of the smallest worker is exactly ten times smaller than the head of the largest. This may not seem much of a disparity; but images taken with a scanning electron microscope can give a different perspective by showing one of the tiniest workers perched on the head of her larger sister, reminiscent of a beetle on the skull of a man.

It must be added that ants come in all shades, though they don't go in for very bright colours. The fact is that, unlike most other members of the animal kingdom, ants have no need of bright colours either for the purpose of blending into the environment and thus escaping from predators or for attracting a mate with a nuptial display. But one does find black, brown, red, and even green or silver ants. There is no such thing as uniformity among ants.

From the underworld to the top of the world

This diversity of ants' appearance and morphology is matched by a wide variation in their ways of life and habitats. There are huge disparities in their housing arrangements, as the French entomologist Pierre-André Latreille observed as long ago as the late eighteenth century: 'The dwellings of ants are as different from one another as the Louvre is from the hut of the Laplander.'

Most species do what the southern wood ants of Europe's forests do: they set up house on dry land. In the tropics, they often live inside small pieces of decomposing wood; and in cooler climates, they protect the colony by building nests.

There are some for whom the open air is almost a foreign element, whose whole existence more or less is spent under the ground. The tiny *Plagiolepis pygamea* of the south of France actually owe their survival to this circumstance: in the early twentieth century, because they were nowhere to be seen, they were able to avoid extermination by the invading Argentine ants which killed off all the other insects that they encountered. There are other species, such as some whose habitat is the tropical forests, which never come down from the treetops, building their nests in the canopy and making their living from it.

In eating habits, too, ants have had to adapt to the local diets on offer. Depending on where they live, they will eat anything and everything. Some, like wood ants, are omnivorous, and will consume indiscriminately whatever meets their mandibles. Others are more particular in matters of diet, such as the carnivores, which include the army ants of Africa and Latin America, well known for their aggressiveness and ready to eat anything that moves, such as insects, earthworms, and even small mammals which happen to get trapped under branches, where they are killed, dismembered, and devoured. Or there are those which, like the southern wood ants, are fond of honeydew, which is made of vegetable sap ingested then excreted by aphids. Harvester ants eat seeds collected in the fields. Ants of the genera *Acromyrmex* and *Atta*, among the most abundant in tropical America, go one better than that, by actually growing their own fungus, their staple diet, in humus or on fresh leaves, which they then cut off.

This review of the diversity of eating habits would be incomplete without a mention of cannibalism. There are numerous species in which the adults and the larvae eat eggs laid by the queen or even by workers. In fact, they are not 'real' eggs that might hatch out as females or males. They are soft; they have a thin skin; and their sole function seems to be to provide the colony with an inexhaustible supply of ready nourishment. The larvae themselves often end up as food for the colony, albeit only in times of shortage. However, there are some primitive species, particularly the *Amblyopone silvestrii*, which use the larvae as the sole source of food for the queens. It should be said that the queens do not actually kill their offspring, being content to suck their 'blood', or rather their hemolymph. This process was first observed by Keiichi Masuko in his laboratory at the University of Tokyo. He saw the queens working their sharp mandibles to pierce the cuticle of the larvae, then swallowing the drop of hemolymph seeping from the wound. He also observed that the wound closed quickly, leaving only a minute scar on the baby ants, which went on developing apparently quite unscathed by their experience.

3
The secrets of success

That the history of ants deserves to be called a success story can surely be seen in the fact that these tiny creatures have not only managed to set foot on all five continents but have overrun them and thoroughly colonized them. By any standards, this is a fine ecological achievement. It can be explained by a single word: sociality.

Ants are not the only gregarious creatures in the world; but few other animal species have achieved their degree of sophistication in social organization or adopted such a strictly hierarchical structure. Colonies, leaving aside some rare exceptions, are founded on the coexistence of two quite separate castes, of different rank, divided by all things except their genealogy. On the one hand, there are the queens, whose sole function is reproduction; and on the other, their daughters, the workers, industrious and generally sterile.

Entomologists have for ever been intrigued by this mode of communal life. Successive generations of specialists have been inspired by it and have coined imagery to make sense of it in the terms of their period. For instance, Pierre-André Latreille (1762–1833), was a priest who, despite being imprisoned under the revolutionary government of Robespierre, was in tune with his times, describing the society of the ants' nest as a 'republic'. This 'prince of entomologists', as he was known to his contemporaries, defined three orders of 'citizens': the males, the queens, and the workers, whom he called 'mules' or 'neuters'. He was also one of the first to realize that the workers were not sexless, but were 'true females, though impotent' and 'condemned to eternal virginity'. Latreille's compassion for these workers is perceptible; he says of them: 'They are helots, whom Nature has burdened with all the most vexing cares of State. Lest the pleasures of love should distract them from the purpose she has set for them, she has forbidden them the enjoyment of its sweets.' Accordingly, his *Essai sur l'histoire des fourmis de France* ('An Essay on the History of French Ants') presents the republic of the ants as a place of inequalities, hard labour, and dreary chastity.

William Morton Wheeler, on the other hand, who unlike Latreille had no experience of the French Revolution, uses a different image. He sees their organization as a mode of 'anarchistic socialism', explaining that 'each individual instinctively fulfils the demands of social life without "guide, overseer or ruler", as Solomon correctly observed, but not without the imitation and suggestion involved in an appreciation of the activities of its fellows.' This idea is akin to the one expressed by Hölldobler and Wilson, who see ants as having a 'highly developed, self-sacrificial colonial existence'. Of this mode of animal communism they say: 'It would appear that socialism really works under some circumstances. Karl Marx just had the wrong species.'

So, are ants republicans, socialist anarchists, or Marxists? Eschewing ideology and anthropomorphism alike, we could say that they have exploited sociality to the full.

United we build

Among ants, there is no such thing as loneliness. Life is all about being part of the group; and the groups, or colonies, may vary considerably, depending on species and geography, 10,000 being a typical workforce. The basic rule, one of cooperation and task-sharing, affords clear advantages to the whole society.

There is strength in numbers and collaboration, the two fundamental conditions which enable ants to change not only their natural milieu but even the climate in which they live their lives. Whereas an insect of a solitary kind may be able to dig a simple hole in the ground to serve as a nest, groups of worker ants manage to construct commodious and comfortable habitats which are perfectly adapted to the environment. Just think of the mound-shaped ants' nests made by wood ants. The shape of these hillocks will vary from place to place, being flattish in parts of the world where temperatures are clement, and more rounded in harsher climates. In this way, the architect workers make the most of the available sunlight, actually making their mounds from dark material so that they function as 'excellent suntraps', according to the Swiss entomologist Daniel Cherix. It is also a way of keeping the temperature inside the nest stable and high, between 22 and 30 degrees Celsius, even when the outside air is much cooler or much hotter.

However, the mound is only the tip of the iceberg so to speak, since much of the nest is under the ground and contains a sizeable network of passageways. There is an intricate system of tunnels connected to more open spaces which function as larders, egg storage rooms or larvae banks, where the conditions of temperature and humidity are maintained at the optimum level.

United we feed and fight

Communal life is also a guarantee of efficacy in the search for food. Minute as each ant is, several of them can work together to capture prey that is much bigger than they are. Six of them can immobilize a large insect, each one holding on to a leg. Even small mammals may succumb to the sheer numbers of ants. The scouts do not hesitate to call up reinforcements when they discover a particularly abundant source of food. By collaborating, the workers can kill and transport to the nest animals which may weigh ten or twenty times more than a single ant, a feat which no individual could accomplish unaided.

Numbers also represent a great advantage if a colony happens to be attacked. Any enemy which manages to breach the fortress sets off a general mobilization. Some of the workers set about repelling the invader, while others make sure of transporting the store of eggs to safety. In joining battle, the workers are fearless, and if need be will even sacrifice themselves for the sake of their society. Any fighter killed on the battlefield is replaced by another, the purpose being to save the queen, whose survival guarantees the future development of the whole colony. Then once the danger is past, all the workers make haste to repair the damage caused.

In all this agitation, the queen herself plays no part. Usually she sits tight in the centre of the nest, in the most protected part. Everything is done with the aim of saving the queen and her brood of eggs, thus ensuring the future security of the colony.

Ease of reproduction

When it comes to reproduction, these conditions of life make for excellent results. Compare this, for instance, to the life of the poor female of the solitary wasp, with all the tasks she has to accomplish before she can procreate. She must first make her nest, then go hunting, and capture a caterpillar, a fly, or a grasshopper to serve as food for her larva-to-be. Only then can she take the time to lay an egg; and she still has to be careful to close up the nest before flying off. For a single female, that makes for a lot of work. And if she makes a mistake at any one of these stages of the process, she will be unable to reproduce.

Queen ants, by contrast, once they have managed to create their colony, generally have to do none of these chores. They lead a pampered life, with nothing to do but lay eggs, letting their daughters deal with all the practical matters of housekeeping. Nor are any of these workers indispensable: there are so many of them that, if any individual fails to fulfil its obligations, there will be plenty of others to carry out the task in its stead. Come what may, the work must go on.

The other great benefit of this cooperation is that it helps the group to cope better with the unforeseeables of life. There is even altruism at work among ants, which sacrifice themselves for the sake of their clan. These are the secrets of their success, as can be seen in their survival rates, which are generally much higher than those of other insects.

Exceptional longevity

Insects for the most part have a short life expectancy. The life of many mayflies lasts for a single day; and most other insects live for only a few weeks. But social insects—ants, termites, bees—can survive for months, if not years.

That sociality and longevity go together makes intuitive sense: to be not only alone and small, but to have few means of self-defence as well, must make for vulnerability. The disparity is in fact much greater than one might suppose. A comparative study of more than 150 different kinds of insect carried out by our team at the University of Lausanne concluded that, on average, ants and termites live 100 times longer than their solitary ancestors, the wasps and cockroaches.

Among ants, however, life expectancy depends on which caste you belong to. If you're a male, for example, you lead a short life.

Your contribution to the life of the colony is so limited that you could almost be ignored: no sooner have you fertilized a young queen, a few weeks after your birth, than you die.

Workers, depending on which species they belong to, can live for two months or even two years. Or rather, they can live that long under the artificial conditions to be enjoyed in a laboratory, where they are protected from any danger. Things are, of course, very different in the wild, where any ants that leave the nest are exposed to parasites and predators or may be crushed underfoot by a pedestrian, and may survive for just a few weeks or even days. Take *Cataglyphis bicolor,* studied in the wild in southern Tunisia by two Swiss entomologists, Paul and Regula Schmid-Hempel. Once they start foraging, the life expectancy of workers is only about six days.

Queens, on the other hand, can live to a ripe old age, ten to fifteen years on average. The oldest ever recorded was a common or garden black ant (*Lasius niger,* a species that lives in urban areas as well as in the countryside), which survived for twenty-eight years in a laboratory. Other individuals have lived almost as long, such as an Australian *Camponotus* which died at the age of twenty-three, or some queens of *Lasius flavus*, small yellow ants found in European pasturelands, which entered the record books by living in captivity for eighteen to twenty-two years. The longevity of queens depends on their way of life and whether or not they have competitors: in colonies where they do not have to share power, they lead a more stable life and die older than in colonies where several queens coexist.

It must be stressed, though, that all queens owe their longevity to the fact that the life they lead is a social one. Kept warm in a cosy nest, spoiled and protected by the workers, queens are safe from accidents, predators, such as other insects, birds, mammals, and humans, and even from parasites. So natural selection has had a vested interest in these individuals, endowing them with physiological mechanisms for the repair of cells or DNA, which would be of little use to insects that die young. This has given them even longer lifespans.

Because they live such a long time, the mothers have plenty of time to produce an abundance of offspring. Even in the least prolific species, the queens give birth, over their whole lives, to several hundred workers, in addition to ten or twenty virgin queens and males. A queen of the leaf-cutter ants of Central and South America may produce 150 million workers, two or three million of which may be alive at any one time. Quite a family! However, the fertility champions are the African army ants, which have double the reproductive rate of the Latin American leaf-cutters, and have more daughters in a lifetime than there are people in the United States.

4
A huge impact on the environment

Size has nothing to do with it. Being tiny does not prevent a creature from having a substantial impact on its environment. This is certainly the case with ants, which, with their population of millions of billions, their nests holding thousands of individuals or many more, can make a real mark on the habitats they colonize. Just by building their nests, they displace such quantities of earth that, taken as a whole, they turn over almost as much as earthworms. In addition, they have so many mouths to feed that, when they mount raids on the plants or invertebrates in their vicinity, they can change the nature of the flora and fauna of some parts of the globe. This is why Patricia Folgarait of the Faculty of Agricultural Sciences at the University of Buenos Aires goes so far as to define ants as 'ecosystem engineers'.

So, are ants engineers? They can certainly construct formidable earthworks in building their nests. In making their networks of subterranean galleries and chambers, they excavate and displace huge amounts of earth. In the deserts of Australia or North America, workers can turn over 400 to 800 kilograms of earth per hectare per year. According to the Australian geologist T. Ron Paton, the record in this activity is ten tons of earth per hectare per year, established by species living in sub-tropical regions, whether humid or temperate. Thus ants, albeit quite unwittingly, bring about considerable changes in the physical properties of the soil, by increasing its porosity and improving drainage and aeration. As diggers and delvers, they may be not quite as good at these jobs as earthworms; but by being more widely distributed on the face of the planet, the total effect of their earth-moving and soil-renewing activity is probably equivalent to that of the worms.

Ants do not only turn over the earth. They also drag scraps of vegetable material into their nests and accumulate refuse and excreta, the effect of which is to alter the chemical properties of the soil. The surroundings of ants' nests are rich in mineral compounds, especially nitrogen, phosphorous, and potassium, as well as in organic matter. In a single chamber created by Brazilian *Atta capiguara* (a space, mind you, that measured 1.5 metres across and five metres high), the amount of organic material found amounted to 500 kilograms. It should be added that, for the purpose of cultivating their fungus, *Atta* bring back to the nest huge quantities of leaves; and in the closed space of the underground chambers this vegetable matter decomposes much more rapidly than it would if left to its own devices. In humid forests, where it is generally difficult for nutrients to penetrate the soil, the leaf-cutting species are in fact largely responsible for the deep fertilization of the ground. Generally speaking, though, the effect of ants' nests on their surroundings is to enrich them with humus. So, indirectly, the 'ecosystem engineers' often have a beneficial effect on agricultural production, especially in grasslands where the soil is naturally poor, as Folgarait has established. As for harvester ants, they contribute to the spread of seeds, since they lose some of their booty along the way to the nest. In the many parts of the world where harvesters occur, from Europe and America to Africa and Australia, they considerably promote the numbers and distribution of flowering plants and they also influence the development of soil bacteria. The American entomologists Sally D. Hacker and Steven D. Gaines state quite categorically that these ants play a part in increasing the earth's biodiversity.

The other side of the coin, however, is that though harvester ants contribute to the spread of seeds, they also consume huge amounts of them. Experiments conducted in the deserts of Arizona have shown that, when the ants were eliminated, the density of annual plants quickly doubled. Similar studies have been made in Australia and, once the ants were removed, the number of young plants multiplied by fifteen. The fungus-growing *Atta*, too, can cause a fair amount of damage, being so greedy for fresh leaves that they are sometimes called defoliating ants. They are endemic in the Americas, and they can build nests that cover a surface of 600 square metres and house several million individuals. So it is not surprising that some of the larger colonies are capable of garnering in hundreds of kilograms of leaves each year, equalling the consumption of a cow. Some colonies can actually cut between one and two tons per year.

Cleaning out the forests

Carnivorous species have huge appetites too. In one season, according to Klaus Horstman of the University of Würzburg (Germany), some of these carnivores can consume up to 40 per cent of the prey to be found in the meadows near their nests. Or, to take an average-sized colony of *Formica polyctena*: over a period of a few months, a colony can consume six million insects and 155 litres of aphid honeydew.

With such ravenous appetites, the carnivores can cause damage by eliminating many useful insects. But, on the other hand, they also promote the good maintenance of forests by ridding the leaves of creatures that eat them. Particularly effective in such clearing operations are the weaver ants which inhabit tropical forests. They are never short of a labour force, as their nests often contain more than a million workers, quite sufficient to protect plantations of coconuts, coffee trees, or eucalypts. In some species, such as *Ectatomma tuberculatum*, though the colonies are smaller, the density of the nests is so great (between 3,000 and 27,000 per hectare) that the effect is similar. These ants manage to take about 260 million prey per hectare per annum, which is a boon to Mexico's coffee plantations. One can understand why such greedy creatures have been recruited for biological combat: as long ago as the third century, the Chinese used weaver ants as forest rangers; and the idea has been revived in the twentieth century, in Germany and Italy, then in Canada, where wood ants' nests have been introduced into forests where there were none.

So wherever they go, ants leave their mark on the surrounding flora and fauna. If they had never existed, various species of insects would probably have disappeared from the surface of the Earth and others would have proliferated. Were it not for ants and the effects they have on plants, the whole process of evaporation would be different and even the climate of the planet might have been altered. Idle though such speculation may be, given that ant colonies cannot be removed from the real world, one thing is certain: if there were no ants, our environment would not be as it is today.

5

A long long story

Extant ants are the result of a very long and ancient history deriving directly from solitary wasps and beginning, quite probably, a good hundred million years ago.

Insects were among the very first creatures to colonize the land, 400 million years ago, according to paleontologists, that is, during the Primary Era. Two hundred million years later, probably during the Jurassic period or at the beginning of the Cretaceous, termites made their appearance. A further hundred million years had to elapse before the first social bees, the wasps, and ants turned up.

Paleontologists began to take an interest in the past lives of ants in the middle years of the nineteenth century. The very first fossils to attract their attention were preserved in amber from the Baltic; and they bore such a close resemblance to modern species that one of the naturalists examining them thought they were hoaxes. This turned out to be untrue, and nowadays it is established that these first fossils do indeed derive from ants that lived in the early Tertiary Era, thirty million years ago.

Myrmecologists suspected that the origin of ants went back much farther than that, a hunch which has now been confirmed. One of the most significant pieces of early evidence turned up, quite by chance, in the United States. In 1966, a retired couple, Mr and Mrs Frei, of Mountainside (New Jersey) happened to find an intriguing fragment of amber enclosing two ants, which they decided to donate to science. After a detour to Princeton, the piece of amber eventually reached Frank M. Carpenter at Harvard. A student of his, Edward O. Wilson, having examined the insects in all possible ways, became convinced that they were primitive worker ants, caught in the fossilized resin of the sequoias which grew at Mountainside about ninety million years ago: 'They had a mosaic of anatomical features found variously in modern ants or in wasps, as well as some that were intermediate between the two groups.' Wilson's excitement, quite understandable, comes through in the adjective he uses to describe the insects' morphology: 'astounding'. He goes on: '[They had] short jaws with only two teeth, like those of wasps.' They also had 'what appears to be the blister-like cover of a metapleural gland, the secretory organ (located on the thorax, or mid-part of the body) that defines modern ants but is unknown in wasps'. In addition, they had 'an ant-like waist, yet one that is simple in form, as though it had only recently evolved'. He was convinced that he had found 'the missing link to the ancestral wasps'. With pride in his discovery, he dubbed it *Sphecomyrma freyi*, the genus name meaning 'wasp-ant' and the species identifier being a compliment to Mr and Mrs Frei.

Not long afterwards, other fossils from the same period were found in eastern and central Siberia, in Khazakstan, and in Alberta. Despite these finds, a shadow of doubt remained, for the *S. freyi* did not in fact appear to possess the 'secretory organ' mentioned by Wilson, the 'metapleural gland' situated at the base of the rear legs which secretes antibiotic substances that protect the insects against bacteria and parasites. It was the absence of the gland, one of the characteristic anatomical features of ants, which led to the suspicion that the fossilized ants did not belong to the Formicidae.

This doubt was eventually cleared up by two entomologists from the American Museum of Natural History in New York, David Grimaldi and Donat Agosti. With the help of fossil collectors and volunteers from the Museum, they were able to work on other specimens of *Sphecomyrma*, also found in New Jersey, which they dated as ninety-two million years old. On three workers and four males, the much-debated gland was visibly present, thus reinstating the Sphecomyrminae as genuine ancestors of ants.

Wasp-ants are very probably not the oldest of these ancestors. More recently other fossils dating from even earlier periods have been discovered. The most ancient of these may well have been identified in 2004 by André Nel and his colleagues at the Musée national d'histoire naturelle in Paris. Their conclusion about a specimen found in Charente-Maritime in western France, caught in amber from the lower Cretaceous, is that it is about 100 million years old, which is why they have called it *Gerontoformica cretacica*. The ant, which is 'almost complete' and without the slightest trace of wings, was 'certainly a worker'. Given the length of its legs, its big strong mandibles, and the shape of its teeth, it seems likely that it was carnivorous. On the other hand, the Paris entomologists acknowledge that 'unfortunately' they cannot state with certainty that their fossil possesses the all-important metapleural gland. Despite this missing feature, Nel is sure that *Gerontoformica cretacica* rightly belongs in the broad family of the Formicidae.

According to Grimaldi and Agosti, it was less than 140 million years ago when ants parted company with their ancestors the wasps, an event which they date more precisely to between 110 and 130 million years. This opinion is widely shared, in particular by André Nel, who points out that 'no ant fossils have ever been found in ambers from Lebanon, which are 130 million years old'. Nel adds, however, that it may be possible that 'ants lived inconspicuous lives for a long time or even that we are incapable of recognizing the oldest specimens'.

The fact is that there is no consensus among scientists on the matter of accurate dating of the period when ants separated from their common ancestor and formed a distinct family of insects. This is a question which has inspired an ongoing debate, as can be seen from two studies, both published in 2006, which come to different conclusions. The first was carried out by a team led by Philip S. Ward of the Department of Entomology and Center for Population Biology (University of California), and including colleagues from the National Museum of Natural History in Washington DC and the California Academy of Sciences (San Francisco). With the aim of putting the competing hypotheses to the test, they compared an impressive amount of genetic data, establishing DNA sequences from 162 species representing the twenty sub-families of ants. These they collated so as to trace the genealogy of the different insects. The conclusion they came to after such a Herculean effort is as follows: 'Our age estimate for the most recent common ancestor of extant ants ranges from around 115 to 135 million years ago.' This coincides with the results of Grimaldi and Agosti, implying that any Jurassic origin of ants is 'highly unlikely'.

One might think that this settles the matter; but one would be mistaken. Using similar methods, a group led by Naomi Pierce of the Museum of Comparative Zoology at Harvard have reached quite different estimates, which push the age of ants back farther into the past: 'Most of the sub-families representing extant ants arose much earlier than previously proposed.' Their origin, it is argued, lies between 140 million and 168 million years ago, at a time somewhere between the Early Cretaceous and the Middle Jurassic.

It is difficult to tell which of these competing findings is more accurate. One thing, however, that now seems certain is that ants existed as long ago as the Cretaceous. It was at that time that they began their spread over the surface of the Earth. To begin with, this was a very tentative process; and so far not many really primitive fossils have come to light, a mere 1 per cent of the traces of insects found in deposits from that period.

With the beginning of the Tertiary Era, sixty-five million years ago, ants started to diversify, as did mammals of course, giving rise to the various sub-families now known to science. Specimens found in sediments dating from the Eocene, about fifty million years old, clearly belong to modern groupings and their anatomy is similar to that of extant ants. All the evidence suggests that ants had contrived by that time to colonize the whole planet and establish their predominance over other living species.

Ancient colonies

While the examination of fossils can teach us much about the anatomical evolution of insects and even about ancestral ants' caste systems, the analysis of extinct species can tell us nothing about their social behaviours. This is why, at the time of Wilson's discovery of the very first *Sphecomyrma*, some of his colleagues took it into their heads to look for the most primitive ant forms, not among fossils, but among living ants.

Not surprisingly, they turned towards Australia, the land of many archaic life-forms; and there, in the 1970s, they came upon the rarity that they were after. It was a large yellow ant with protruding black eyes and long mandibles shaped rather like the serrated blades of pinking shears, bearing the name *Nothomyrmecia macrops*. This species had actually been described in the early 1930s, by an expedition that set out from the west and headed south, on a hunt for unknown insects. Having trekked through eucalyptus forest and across sandy wastes, among the booty they brought back were two large yellow ants, which they donated to the National Museum of Victoria in Melbourne. The resident myrmecologist, John Clark, realized that they belonged to a new species of a new genus. He it was who baptized them *Nothomyrmecia macrops* (they are also sometimes called 'living fossils') and preserved them in alcohol as part of the display at the museum.

There was, however, a problem: the insect hunters who had discovered these weird specimens had not thought to record the exact location where they found them. So, twenty years later when entomologists realized the scientific value of *Nothomyrmecia* and its place in the history of ant evolution, the hunt had to begin all over again. Although the hunters retraced the steps of the original expedition, they came back empty-handed. But this was only the beginning of a search which was to go on for many years.

It was an Australian, Robert Taylor, who eventually found the elusive insect, quite by chance, in 1977. It is said that, one night, as he was relieving himself, the beam of light from his head torch picked out a *Nothomyrmecia* worker crawling up a tree-trunk. This was a great surprise to him, as he had just stumbled upon the primitive ant in the township of Poochera (South Australia), more than 1,000 kilometres away from where he thought he might find it. Since then, Poochera has become something of a sacred site for myrmecologists, who come from all parts of the world to see for themselves the colonies in the wild and to study every last detail of their lives; and *Nothomyrmecia macrops* is now very well documented in the scientific literature.

Just like modern ants, they live in groups, but the colonies are small, never containing more than about 100 individuals. They even have a hierarchical structure, with a reproductive queen and her sterile daughters; and the workers have some of the habits of more evolved species, such as reciprocal grooming. Suffice to say that, though these creatures do have a social organization, it is still rather rudimentary. There are even ants which are more primitive than the *Nothomyrmecia* discovered by Taylor, of which the *Amblyopone* are considered to be the most archaic. So, even though *Nothomyrmecia* aroused immense interest, it must be admitted that they were not actually the very first ants to colonize the earth.

Part II
Social Life

Drawing 2 Parasites Queens of *Teleutomyrmex schneideri* have no workers and live in complete dependence on queens of *Tetramorium caespitum*. They cling to their host's abdomen or thorax and live there permanently.

6
The birth of the colony

The great day arrives and the nests are teeming with activity and excitement: the virgin queens and the males are ready for their mating flight. As Réaumur said in 1731: 'The wedding ceremony of the ants must be celebrated in the air.' The young queen has been an adult for the last two weeks, during which time she has been accumulating such reserves of fat that, since becoming an adult, her weight has more than doubled. She is now ripe for the mating flight.

Females and males alike await the most favourable moment, on the watch for clement weather conditions. For a sortie into the outside world, mild weather is best, since if it is too chilly these cold-blooded creatures have difficulty in activating their muscles and beating their wings. A slight wind is also desirable, but not too much, for these insects are not well adapted for flight. In addition, the optimum time for the future queen to make her exit is just before or just after rain, because then the soil will have softened and this will make it less difficult for her to dig her nest.

How the inhabitants of the nest contrive to gauge the atmospheric conditions is not properly known. We do know, however, that the ants react suddenly, as though in response to a signal. At a precise instant during this particular day, they become very active and start to bustle about. The workers leave the nest and patrol the environs. To the eye of the experienced entomologist, it is clear from this behaviour that the departure is imminent. Then suddenly the males fly off, followed about twenty minutes later by the females, the twin operations being perfectly synchronized. In grasslands where there are several nests, the departures happen simultaneously.

In some species, the females do not fly far, but stay close to the nest. Some of them even stay on the ground where they emit chemical signals, the sexual pheromones, to attract their partners. In some other species, however, the young queens climb to about a hundred feet, and the most adventurous go even higher to find a partner. Some mate with a single male, some with several or, in certain species, even more than ten.

The males are lucky to mate with any female; and if they succeed, there is little likelihood that they will be able to couple more than once, their reserves of sperm, formed when they were still in the larval stage, being limited, because as they reach adulthood, their testes completely degenerate.

In any case, neither those who do manage to reproduce, nor the unlucky ones who failed to mate, will survive their nuptial ballet. More often than not, they are unable to feed. The sugar reserve which they have accumulated during their time in the nest enables them to fly about for an hour at the most. They then fall to the ground exhausted, as though they have run out of petrol. Some ants do this in huge numbers, for example those living in eastern regions of the United States, where the mass demise of the unsuccessful candidates for reproduction can be witnessed every year at the end of the summer. This is the season of the clouds of *Lasius neoniger* that Americans call 'Labor Day ants', which are always reproductively active in the days just before or after the public holiday on the first Monday in September. Once they have fallen to the ground, the bodies of the males do not lie about for long, being soon found by predators, many of whom are other ants, which make short work of them.

Some males meet an even more grisly end by being killed by the female they are in the act of mating with. This sad fate awaits the male members of *Dinoponera quadriceps*, a quite remarkable Brazilian ant. There being no queens in their colonies, at the mating season the young workers engage in ritualized fighting to determine which of them has the right to reproduce. The aggressiveness of the winner is not in any way diminished by her becoming the dominant female. As soon as she is fertilized, she goes back to the nest, dragging the still coupled male behind her like mere prey. Once she is inside the nest, and still in the act of mating, she puts a sudden end to him by cutting him off at the abdomen. His genital parts remain inserted in her own reproductive apparatus; but she will manage to rid herself of them, though it may take her half an hour.

Being male in the matriarchal world of ants is definitely not an enviable role to play. As Wheeler says: 'The males are in every sense the *sexus sequior*,' which is why he compares the societies of the ant world to 'certain mythical human societies like the Amazons'. The males are mere sperm factories, surviving only for as long as it takes to donate their semen and thus ensure the transmission of their genes.

Sperm bank

Once fertilized, the young female goes on her way; she pulls off her own wings for which she has no further use as she will never mate again. The sperm acquired during her mating flight she keeps in an oval pouch located in her abdomen, known as a spermatheca. In this way, she accumulates a stock of several hundred thousand spermatozoa, sometimes even several million, in a dormant state. For the rest of her life, she will draw upon them to fertilize her ova, eventually giving birth to offspring whose fathers have been dead for many years. The survival of such a natural sperm bank is somewhat mysterious. How do queens manage to keep spermatozoa in their required functional state? It has been suggested—though this remains a hypothesis—that glands close to the spermatheca secrete a substance capable of nourishing the male cells.

Be that as it may, the queens are now primed and ready to found their family. This they do in ways which vary from species to species: the fertilized females of wood ants, for instance, are incapable of establishing their own colony; but they have invented a subterfuge, in that they act as parasites on established colonies, sometimes of another species. The behaviour of others, such as the young queens of the army ant, is even less complicated: having no wings, they never leave the nest; they merely excrete their pheromones and wait for male outsiders to come in and mate with them. If two queens are fertilized in a nest, the colony splits in two.

These wood ants and army ants, however, are exceptions. In the majority of cases, the young fertilized female initiates a new colony without any help (claustral colony founding) after the mating flight. Having to raise her brood entirely from the limited energy obtained from the histolysis of her wing muscles and the fat that she accumulated before the mating flight, she first produces very small workers so as to increase their numbers and thus the chances of success. It is better to go for numbers and to have about twenty undersized offspring than to have larger but fewer workers. By the following year, this first generation has reached adulthood and the queen, surrounded by her earliest workers, which busy themselves with the mundane household chores and the hunt for food, can now quietly devote herself to her primary function—that is, she lays. The foundations of a society are in place and the colony can now develop properly. For three or four years, it will grow exponentially until it reaches its maximum size.

Once the group is well established, the time has come for the queen to start conceiving not just workers, but also males and fertile daughters. Among ants, sexual differentiation is a simple process: any unfertilized egg (a haploid) will produce a male, whereas any fertilized egg (a diploid) will develop into a female. One of the most surprising aspects of the whole business is that the queen appears to have the power of choosing the sex of her young, of 'deciding' whether or not to fertilize an egg by opening or failing to open the aperture of the spermatheca. The 'choices' she makes vary in accordance with the age of the colony; and in the early days, she will produce only workers and no males. Her attitude also depends on the time of year: a great number of her earliest eggs, laid in the spring, will be males, and later on she will fertilize almost all her eggs so as to have mainly daughters.

The destiny of daughters

Whereas the existence of males is determined by the non-fertilization of the eggs, the future of females is not the outcome of solely genetic factors, since queens and workers do not differ genetically. The social environment, in fact, has an important role to play in their development. The question arises, though: which of them is it, the queen or the workers, who determines the destiny of the daughters? For a long time it was thought that this, too, depended on a choice made by the queen; and it is a fact that, in several species, the mother ants do secrete pheromones which inhibit the production of new queens. But if there was any 'chemical manipulation', the offspring might develop a resistance to it. The chemical emissions are really signals given by the queen to the whole colony, as though she is saying, 'See how fertile I am—if you help me, you will have many brothers and sisters.' In fact, it is the workers who largely control the destiny of the larvae, by varying the quantity and quality of the nourishment they give them. This is what favours the hatching of queens or workers.

In this way, each colony will produce between ten and 10,000 new queens, each of which will fly away in her turn. These young virgins, however, will then have to compete vigorously with young queens from other nests to find a mate and start a new colony. By the end of the whole process, very few of them will have survived and succeeded in founding their own society.

7
Division of labour

Once the new colony is established and the queen has managed to get through her first year unaided, she can then devote herself to her essential mission. She turns into a real egg-laying factory, with the single job of ensuring the reproductive future of the family. All the other jobs, the ones on which depend the survival and the well-being of the clan, are done by the workers. Demarcation never leads to dispute, each denizen of the nest, queen or worker, having its designated function in an efficient division of labour.

Dyed in the wool republican that he was, Pierre-André Latreille was offended by the organization of such societies, from which, he said, 'equality seems to have been banished'. He was full of indignation at the fate of the workers and the 'burden' imposed upon them. But he was glad to be able to add that, 'Compensation comes about in the whole. For authority, power, and strength reside essentially in these tiny beings whose fate seems so unenviable to us. They it is who are the providers and the guardians of a family in its cradle. The existence of future generations is entrusted to their care.' His anthropomorphism showing, he added: 'Rearing these adopted children no doubt affords them true happiness; and the pleasures they derive from playing this part in mothering outweigh their exclusion from others.' Between Latreille's view and our own time, Darwinism has intervened and we now know that the causes of the workers' altruism are in fact much more prosaic.

Nevertheless, if we leave aside the moralizing that accompanied Latreille's view of ants, it must be admitted that his observation was in large measure accurate and that, in what he saw as 'the republics of the ants', a way of apportioning tasks has developed which can be seen to be very effective.

The nursemaids

One of the tasks which fall to the workers is to look after the brood, composed of eggs, larvae, and pupae, the three separate phases of maturation that ants go through before reaching adulthood. Once an egg is laid, its cells divide over the course of one to two weeks before it transforms into a larva, which will grow and develop. But it is not until some two to six weeks later that these simple organisms, which have a digestive tube but no appendage of any kind, turn into pupae and gradually acquire the morphological features—eyes, legs, antennae, etc.—which are those of the adult they will become. In *Formica* or in the large European *Camponotus* ants, the pupae are generally enclosed in a pupal case, which gives them something of the appearance of silkworms. With or without pupal case, however, they are immobile and, if left to their own devices, could never survive—which is where the nursemaids come in.

Very attentive nursemaids they are too. For example, in soil-dwelling ants, they keep moving their charges from place to place, several times a day: in the morning, they carry them up to the surface, where it is warmest; and later on, when the heat outside is becoming too much for the brood, they carry them down again to the cooler underground passageways of the nest. They go to great lengths to keep the conditions of warmth and humidity at the levels most conducive to the proper development of the brood. During heatwaves or times of drought, the workers form into regular brigades, busying themselves in all directions, conveying water from mouth to mouth and regurgitating it on to the walls and floors of the nest.

The nursemaids also attend to hygiene by cleaning the eggs and using their own fungicidal or antibiotic secretions to kill parasites. In addition, with the purpose of restricting the development of pathogens, they take care to lodge the brood in areas of the nest that are well away from where the food is stored. This is also why they so seldom leave the nest, since during each foray into the outside world they risk being contaminated.

In the Jura region of Switzerland, the ant *Formica paralugubris* go even farther by becoming pharmacists. As has been demonstrated by Philippe Christe and Michel Chapuisat of the University of Lausanne, they go in search of spruce resin, which has been shown to have anti-microbial properties. By distributing the resin in the nest, they can decrease the number of pathogens. The use of medicinal plants is of course widely practised by certain animals; among other species, bears, chimpanzees, and starlings go in for self-medication. But in those cases, the behaviour is a strictly individual thing, quite different from the collective endeavours of worker ants, who are acting for the good of the entire colony. It seems that they make a point of using the resin preventively and do not wait until the nest is infected before setting out on a quest for the pharmaceutical substance. In some large ant-mounds, entomologists have found up to twenty kilograms of resin.

The nursemaids leave it to their sisters the foragers to seek out the food, though it is they themselves who will give it to the larvae (the eggs and the pupae not needing to be fed). It is true that in some primitive species, the larvae feed themselves; but such species are relatively rare. In general, the nursemaids behave like mothers mashing up food for their babies. In what can be called their 'social crop', they premasticate the food, predigest it, then regurgitate it as tenderized nourishment for their brothers and sisters.

The development of the larvae is also influenced by how the workers treat them. In particular, they take special care of the future queens, giving them more to eat than the other larvae and allotting them the warmest spots in the nest. Similarly, the nursemaids give larger helpings to some of their sisters who are thus intended to grow up as warriors. In these ways, in accordance with the needs of the colony, they determine which caste the larvae will eventually belong to.

The foragers

Foragers are specialists in seeking out food and supplying the colony. In accordance with the ways of their particular species, workers of this caste go out in search of seeds or on the hunt for insects and other small animals. This work of prey-seeking may take them as far as 100 metres from the nest, a practice in which they share out the labour according to size, the largest ants being the ones which venture farthest.

When they locate a source of plentiful food, they will to and fro as often as necessary to exploit it to the full. They never lose their way, thanks to visual landmarks and chemical traces which, like Tom Thumb, they leave wherever they go. At the end of the winter, if the natural markers or their own traces have disappeared, this is not a problem, for, like the wood ants, they will have kept a memory of which direction to follow so as to find good places for hunting or gathering. Things are not so simple for desert ants such as *Cataglyphis*, which cannot rely on any chemical traces they may have deposited, for these will have been blown away by the wind and covered up by sand. However, these ants have found how to cope with this: to orientate themselves, they use polarized light from the sky; and they also contrive to memorize their own movements, the direction to take, and the distances they have covered—they can be seen making a straight line back to their nest without the slightest detour.

This hunting for food always entails an advanced degree of cooperation. The scouts are the first to set off, their job being to explore the terrain and to recruit reinforcements whenever they come across a source of nourishment. Their role does seem to be crucial in supplying the colony. In the *Pogonomyrmex barbatus* red harvester ant, the scouts even separate into two distinct groups, as has been observed by Deborah M. Gordon of Stanford University: the first group, whom she calls 'nest mound patrollers', are outside for no more than a few minutes and do nothing other than scout about the immediate environs of the nest; they are relieved by the second group, the 'trail patrollers', who venture much farther in the search for food. The forager regiment does not set out until these patrollers have returned. Gordon not only observed this sequence of events; she also decided to intervene in this neat choreography by capturing the returning workers. Her conclusions are unambiguous: when the first group of patrollers fail to return, those awaiting them stay in the nest and the seed supply of the colony decreases to zero. If some of the scouts of the second group go missing, the outside activity of the colony is greatly reduced; whereas, if it is the foragers themselves who are removed, this makes for only a temporary hiatus in the harvest work, which proceeds at not much less than the usual rate. Gordon says that 'the nest mound patrollers may assess humidity and temperature'. We might say they are testing the ground, by way of making sure that the weather is good for harvesting. If they do not return, their fellow workers take this to mean it is not a day for being out and about.

Different species have different ways of doing things and work within different constraints. In fungus-growing ants, the large foragers sallying forth to cut leaves surround themselves with 'body-guards'. The fact is they are in danger of death from parasitical flies which lay their eggs on the workers' heads; as the fly larvae develop, they feed on the muscles inside the ants' heads, which proves fatal. This is why, while the larger foragers are busy cutting leaves, small workers are on the watch nearby. Being less bulky and able to move faster, they perch on top of the leaves and attack any winged insects which might come too close to their sisters' heads.

Within the hunting activities proper, the dividing up of tasks may go even further. In some species, such as *Cataglyphis bicolor*, some foragers tend to specialize in tending homopterans, while others preferentially forage for arthropod prey. In some other species, such as *Allomerus decemarticulatus*, workers do not even have to leave their nest to go on a hunt. They just use the plant on which they have established their living quarters to actually make traps in which they catch prey. This particularly ingenious technique was recently described by Jérôme Orivel and his colleagues from the French CNRS (Centre national de la recherche scientifique) and the University of Toulouse. Just imagine: the workers take hair from the host plant (*Hirtella physophora*) and bind it together to form a gallery, its pillars being the follicles left uncut on the plant. This constitutes the base of the trap. The ants then mix into it chewed up remains of organic matter which they regurgitate and shape, next consolidating this mortar with a particular fungus whose filaments, as they grow, will cement the whole thing together. In its finished state, the trap has the form of a sort of tunnel full of holes. The *Allomerus* position themselves inside, at these apertures, with their mandibles wide open, and there they wait. As soon as an insect lands on the trap, they grab it by the legs or by the ends of its antennae and pull. Once the prey is immobilized, held by the legs, other workers emerge from the tunnel to sting it and paralyse it with their venom. All that remains is for the hunters to dismember their catch and carry it off to the pockets of leaves where they live.

This is indeed most ingenious as a hunting technique; but it is also surprising, in that hitherto the collective construction of traps for prey had been seen as limited to the social spiders with their webs—nothing like it had ever been observed among ants. Not only that but, unlike spiders who spin their own silk, these ants show they are able to select from their environment the materials required for the building of such a complex structure. At any rate, this technique perfected by *Allomerus decemarticulatus* is remarkably effective: with their traps, these tiny ants only two millimetres long can capture insects that are more than three centimetres in length and that weigh more than 1,500 times as much as they do.

The builders

Workers act also as architects and builders whose job it is to construct the nest, in accordance with the specifications of volume and structure required by each species and consistent with the size of their colony. Ants will use for housing anything and everything, from mere bits of wood, barely modified, to the most sophisticated nests built up in the trees. Underground nests vary from the simple dug-out about twenty centimetres below the surface, housing a few dozen individuals, to veritable fortresses comprising numerous passageways and many different rooms, which may lie four metres down. The mounds, too, like those raised by *Formica* in the forests of Europe, are much more than just heaps of earth left over from the excavation of a subterranean nest: the surface is often covered in pine twigs, sometimes in pebbles and pieces of charred wood; while inside there is a dense network of tunnels and interconnected chambers, which makes each mound a metropolis.

To dig or erect such edifices, the workers clearly have to help one another and share out the labour. Generally, one group will go and forage for the necessary materials, which they will bring back and deposit unsorted on the building site, letting other workers get on with the finer points of making the dwelling. Nor does the work come to an end once the nest is built. The workers have to enlarge it so as to accommodate the increasing population, just as they have to attend to maintenance and repairs. This is one of the very first activities that wood ants undertake on reawaking in the spring, especially if they live in regions where heavy snowfalls over the winter months may have slightly compacted the nest. According to Daniel Cherix, 'The workers will refurbish the upper parts of the nest by moving, in a more or less orderly fashion, the materials that it's made of.' If the structure has been damaged by predators, the workers carry out the necessary repairs, using the larger twigs as a 'sort of roof framework' on the inside and 'shifting the smaller woodwork, such as pine or spruce needles, to the top of the dome, setting them out in a layer several centimetres deep which will waterproof the nest'.

So much for the sedentary species. There are others, such as Argentine ants, which are more sensitive to variations in their environment and will change habitat several times a year. They move house generally in spring and autumn, endeavouring on each occasion to find locations that will offer the colony the conditions of temperature and humidity most favourable to its continuing well-being. Nomadic as they are, they build nests which are less elaborate than those of their sedentary cousins.

8
Let slip the ants of war

Ants are not only hard-working, attentive, and even altruistic when they stay quietly in the nest taking care of the brood; they have another hidden side to their character. The aggression they are capable of is almost without parallel in the rest of the animal kingdom. They can wage war on the members of a rival colony or when faced with individuals belonging to a different species.

The battles they have to fight arise most often from territorial conflict, whether a colony tries to defend its territory against intruders or to expand it so as to increase the resources available to it. The war effort, however, must be kept proportionate to the benefits deriving from it, particularly any gains in food supply, which must outweigh the losses incurred, as measured by the number of individuals killed in action. Depending on species and the size of their colonies, ants can adopt various different strategies for defence or attack, ranging from full-scale battles to duels or even suicide attacks.

In temperate climates, it is especially in the spring, a time when the brood is developing and insect prey is rare, that ants' warlike impulse revives. It takes less than an hour for two opposing sides to be on a war footing, though both are careful, in summoning their worker-troops, not to enlist more than a fifth of the population of their colony, in case a third clan should take advantage of the deserted citadel and occupy it. The soldiers then join battle, but in a way which will not lead to the annihilation of one of the armies. Since each individual worker is able to recognize the smell of her own cohorts, she can note how many times she touches an enemy and in this way estimate how many of them there are. If one side notices that it is in a position of numerical inferiority, it will retreat a little and abandon part of its territory. Battles usually last for no more than a few hours, given that it is futile for the losing side to fight on and that the winners have no interest in acquiring more territory than they could easily exploit and defend.

Such engagements require proper soldiers, some species even having a special military caste which is morphologically different from the others. European colonies of *Pheidole pallidula* have soldiers which are not only larger than the other workers but have bigger heads. It should not be assumed that this means they have bigger brains making them more 'intelligent' than their sisters; they just have more highly developed muscles, which gives them greater power in the use of their mandibles. The whole military culture of this species is also remarkable. Any colony that becomes aware of another nest in its vicinity can prepare for warfare by doubling the numbers of its troops. Working with the team of Luc Passera from the CNRS's Research Centre on Animal Cognition and the Université Paul-Sabatier in Toulouse, we demonstrated this singular war effort through experimental work on forty colonies which we transported to the laboratory. With great patience, we removed all the soldiers; then we arranged the colonies in pairs in such a way that the sole point of contact between them was a tunnel located near their respective food supplies. Some of the tunnels were obstructed by mesh so fine that only the workers' legs or antennae could penetrate it; and others were completely closed off by plastic film. By the end of seven weeks, the situation was clear: colonies which had been able to make contact with their foreign neighbours had greatly increased their production of soldier larvae; on average, there were twice as many soldiers as in the colonies where contact with the enemy had been prevented. This means that, in response to a presence of other ants perceived as a threat, the nursemaids can vary the feeding of the larvae so as to foster a greater number of soldiers. Since the biomass of a colony, that is the total weight of all the ants taken together, does not change under these wartime conditions, this means that the war effort has a detrimental effect on the development of workers belonging to other castes.

Among *Pheidole dentata*, cousins of *Pheidole pallidula* living in the American south, the soldiers' mandibles are triangular and sharp. Unlike the great majority of ants, they do not use them just for biting the enemy. These aptly named *dentata* use theirs like shears, snipping off the heads and legs of their enemies, or cutting them in two. As weaponry goes, this is formidable, enabling the *dentata* to hold their own even against the red imported fire ant, which lives in the same regions but in colonies that may be 100 times larger than theirs.

Formica polyctena, a species of wood ant inhabiting forests in northern Europe, can turn particularly nasty if food runs short, especially in early spring when growth spurts start within colonies. They mount full-scale expeditions against rival societies of their own species and will also attack ants of other species with such ferocity that they can at times completely rid their locality of the enemy.

Stinging

To protect themselves against predators, defend their territory, and also to capture their prey, many species of ants are equipped with a singularly effective weapon: a sting, which can be used against an enemy for injecting venom. Some Ponerine ants, such as *Pachycondyla tridentata* from South-east Asia, have a way of adapting their strategies to the size of an adversary. Faced with a relatively large arthropod, they will sting it; but if they are attacked by small *Pheidole* ants, they behave quite differently. From the rear of their abdomen they produce filaments of venom which entirely cover the attackers; the latter are bogged down in this frothy substance and instantly lose all their aggressiveness.

In their warlike frenzy, ants will go to any lengths to defend their colony, even including suicide attacks. The outright winner in this department has to be a species of *Camponotus* from the virgin forests of Malaysia. Their anatomy is highly peculiar in a way that makes them walking chemical bombs: they have two huge glands which go right through their bodies and contain toxic substances. If they get into difficulties, they violently flex their abdominal muscles, bursting their cuticle, that is their outer covering, and spray their enemies with deadly venom.

And then there are warriors who never attack but limit themselves to purely defensive strategies. This can be seen especially in *Cephalotes*, a tree-dwelling species which have acquired a veritable suit of armour, with very hard bodies, covered in spikes, and antennae protected by frontal carinae.

Other ants also avoid physical combat if faced with colonies belonging to their own species. Instead of killing or wounding their enemies, they go in for much more peaceable strategies. For example, *Myrmecocystus*, honeypot ants, engage in what can be seen as more or less ritualized tournaments, requiring little violence and with the main objective of merely intimidating or repelling the enemy. The workers scurry about, swelling their abdomens and stretching their legs, as though trying to look larger than they are, an illusion they foster by perching on top of pebbles or higher ground. When two adversaries confront each other, they wield their antennae or their legs, jousting for just the few seconds that it takes for one of them to give up, whereupon they both set off looking for another member of the opposing side. These honeypot ants can also behave like highwaymen, stealing the food gathered by others rather than trying to add to their own food-producing territory. Their victims are ants of the *Pogonomyrmex* genus, harvesters who may pick up termites along with seeds. If the honeypot ants come across them, they examine their booty and, if they find an insect among it, make off with it.

Sometimes, though, the boot is on the other foot, and *Myrmecocystus* will be victimized by some other species. In the deserts of Arizona, the minute *Forelius pruinosus* ants, despite being much smaller than honeypot ants, use their toxic secretions to intimidate them and seize their reserves. According to Hölldobler and Wilson, 'They also occasionally prevent the honeypot ants from leaving the nest altogether by gathering in hordes at the nest holes and using their chemical weapons to drive the big ants underground. The honeypot ants are thus cleared from the hunting areas around the nests, allowing the *Forelius* to harvest a larger share of the available food.'

Looting is in fact far from exceptional behaviour among ants. The European *Solenopsis fugax*, for instance, are also practitioners, and on a large scale. Their workers dig out a network of tunnels joining their nest to the one they intend to invade. Once they have succeeded in penetrating it, they fight off the defenders with repellants that they secrete and, having seized the brood, they carry them away for feeding on later. One of their cunning precautions is to make the escape passage too narrow for the dispossessed workers to follow them, so they can make off with the booty without fear of reprisals. Ants know all the tricks, including some truly violent variants, when it comes to looking after the interests of their own colony.

9
Flexible work arrangements

Ant society, with its nursemaids, foragers, builders, and warriors, can appear to be organized in accordance with the strictest rules of time-and-motion study. This does not mean, though, that workers acquire a single specialized role and keep it for ever. On the contrary, they are quite able to be flexible in their work arrangements.

As a rule, labour is allocated according to the size and age of workers, not all members of the working classes being of identical design. The amount of nourishment given to them by their nursemaids when they were larvae will have determined whether they are larger or smaller. Generally speaking, between 'majors' and 'minors', as they are called, there are differences of weight of between 10 and 20 per cent, though in some species the disparity can be much more marked, some majors being up to 100 times bulkier than the minors. These size differences are more pronounced in certain parts of the body; and sometimes they are accompanied by morphological differences.

In colonies which are populous enough to have a standing army, the warrior ants have larger heads than the others as well as stronger mandibles. Strong mandibles are also a feature of harvester ants, substantial jaws being a prerequisite for crushing seeds. But there are even more surprising disparities, such as a caste of workers among European *Camponotus* whose peculiarity it is to have flat-fronted heads. These so-called 'truncated' ants are unbeatable when it comes to defending their honey against would-be predators: faced with such a danger, several of them will stand together and insert their heads into the entrance to the nest like a stopper, protecting their precious reserve against intruders.

It is logical that the largest workers, whether their heads are of a special shape or not, should take on tasks that require them to leave the nest, to seek out food, for example, or defend the territory, and that the smaller ones should be restricted to domestic chores. Not that size is the only thing that counts in this; age, too, can be important, and the most dangerous work is generally entrusted to the oldest ants. Faced with an emergency, this is a way for a colony to husband its most precious resources.

Room for manœuvre

These rules are not actually inflexible; workers do have some room for manœuvre, enabling them to adapt to different situations. If all the outside workers in a colony are removed, in very short order their sisters inside change their behaviour and come out of the nest to replace the missing foragers or soldiers.

What this means is that any individual worker is able to take on any task, depending on the needs of the colony. If a society is to function properly, 'human resources' must be well managed, so that at any given moment there is a sufficient number of workers capable of doing whatever task need to be done. The question of how this reallocation of resources is effected has long puzzled myrmecologists. For there is neither a central authority at work here nor any hierarchy among the workers. Work arrangements depend entirely on individual initiatives within a system of self-organization which entomologists explain by what they call 'response-threshold model'.

Just as humans do not all have the same sensitivity to pain or stress, not all ants have the same threshold of response to a given stimulus. If for example supplies run out in a nest, it is the individuals who are most sensitive to hunger, whose tolerance in that area is lowest, who are the first to set off in search of food, which they will bring back to share with nestmates, in a process known as trophallaxis. These nestmates will not follow them outside unless the shortage continues. The level of each ant's threshold of hunger is a function of several variables, depending not only on a worker's membership of the major or minor castes and on its age but also on a genetic component. In species where the queen mates with several males, for instance, workers who have a common father will have similar reactions and are therefore likely to be on the same 'shifts' in the colony.

Individual thresholds of response being judiciously distributed throughout the colony, division of labour follows a pattern that we could call 'self-managing'. Whatever the task is, it is done by exactly the right number of workers, neither more nor less. In addition, unlike what happens in hierarchical systems, such as those of primates, say, in which the loss of the dominant individual can cause major disruption to the whole society, worker ants are interchangeable with one another. This is a highly functional structure, making for colonies which are particularly good at coping with adversity.

The workers' elite

Despite self-management and such flexible working arrangements, some workers are more active and efficient than others, a state of affairs that has nothing to do with their morphology or their age. If colonies of *Tapinoma erraticum* are obliged to move house, it will be noticed that some ants, apparently quite indistinguishable from the others, are much more efficient when it comes to transporting the pupae. This is why two entomologists, Simon Robson and James Traniello, have put forward the theory of 'key individuals' whose role within the colony is to stimulate the zeal of the other workers, at which they are so successful that they can even make inactive ones start working. Robson and Traniello report that when a *Formica schaufussi* forager worker acting as a 'scout' locates a prey too large to be retrieved individually, it organizes cooperative prey transport by recruiting nestmates. During this process, the scout plays a key role in maintaining the cohesion of the retrieval group. Indeed, when the scout is experimentally removed, the recruited workers composing the retrieval group typically abandon the prey and cooperative foraging is terminated.

What is it that makes some individuals in a colony become model workers, forming what the French biologist Marguerite Combes, who discovered this phenomenon in 1937, called 'the workers' elite'? Is it caused by genetic factors, individual experience, the social context, or by mere chance? For the time being, we cannot answer these questions.

Two things, however, are certain: the sharing of tasks is universal among ants; and this phenomenon, which can be observed even in the archaic species, appeared very early in their history. Even in colonies of the dinosaur ant *Nothomyrmecia macrops*, there are several individuals whose special job it is to stand guard at the entrance to the nest. This is admittedly a very rudimentary mode of allocating work; but it represents the inception of a process that evolution was to transform into the much more elaborate division of labour of more derived ants. These intricate work-sharing systems, seldom matched by any other living creatures, are undoubtedly one of the main keys to the remarkable ecological success of ants. It must be said, too, that ant societies possess particularly sophisticated systems of communication.

10
Communication systems

Human beings, who tend to see communication as inseparable from language, need to be reminded that exchanging information is not just a human thing, but is widely practised by many other animals. Indeed, the very survival of many species depends on it. Without communication, sexual reproduction could quite simply not exist, since it is through the sending of specific signals that an individual indicates to its fellows which sex it belongs to and thus attracts the partner with whom it will mate. Animals are like humans in that the higher the level of organization in a society, the more complex is its communication system, because its members must be able to exchange information that is more copious, accurate, and elaborate. And ants are no exception to this general rule.

Being the most social of social insects and living in a regime based on mutual assistance and task-sharing, over time ants have become real communicating machines. Like most other animals, they have achieved this through the use of a chemical language in which the 'words' are in the form of pheromones, substances detectable by smell and by taste which can be produced and recognized by all individuals. Ants are walking bundles of secretory glands (they have about forty of them, mainly in the abdomen, the head, and the legs), which enable them, depending on species, to emit between ten and twenty different pheromones, each of which has its own 'meaning'. Some of these pheromones have a marked sexual connotation, such as the ones used by virgin queens for attracting males. Some are produced by workers for recruiting their sisters or for alerting them to danger. Others are used for marking territory, for identifying members of their colony or conversely for detecting foreigners. There are also royal pheromones, which are the source of the strong attractiveness of queens for their 'court'.

Such chemical communication considerably increases the efficacy of the work required to keep the colony constantly provisioned. A scout finding a plentiful source of food not only carries back a sample to the nest; she summons assistance from her sisters, so that the harvesting or hunting will be more productive. This may not always have been the case: workers of the primitive Australian *Nothomyrmecia macrops* go hunting for insects on their own. In other species, recruiting is individualized: an ant that has found food returns to the nest and enlists the assistance of a single worker who will follow her back out. At each exit and re-entry, the number of foragers will double; but such a rate of recruitment makes the supplying of the colony time-consuming and laborious.

Accelerated recruitment

More 'modern' ants systematically use pheromones to make things move more quickly and to recruit many workers at once. On the way out, their scouts leave a chemical trail behind them so as to find their way back. If they find a source of nourishment they secrete even more pheromones on their way back, so that the other workers have a clear path to follow.

The behaviour of the scouts differs according to how large the prey is, as shown by observation of the Mediterranean *Pheidole pallidula*, a species which feeds on seeds and insects of very diverse sizes. These ants offer a good model for the study of what Jean-Louis Deneubourg, an entomologist at ULB (the Université libre of Brussels), calls the 'general dynamics of recruitment'. When a scout comes upon a fly or other insect small enough for her to carry unaided, she takes it and brings it back to the nest without bothering her nestmates. But if the prey, a cockroach, say, is too big for her, she goes and recruits other helpers. The recruiting process works rapidly, for within an hour there are 250 ants working away at the prey, most of which are minors, though there are also a few majors that have been drummed up to help with cutting up the bodies of the insects, which are then taken back to the nest in small pieces.

As long as some food remains, the workers keep toing and froing, leaving their chemical trace each time. The effect of this is to increase the concentration of pheromones laid down, which makes the path easier to follow and also attracts ants whose response thresholds are higher. It is this phenomenon of reinforcing the signal which explains why, when ants have a choice among several different paths leading to a food supply, they always take the shortest. The Brussels team of Deneubourg, Serge Aron, and other colleagues produced proof of this in laboratory experiments with Argentine ants. Having set out two sources of food at slightly different distances from the nest, they observed that the workers tended to choose the nearer one. A moment's thought shows why there is nothing surprising in this: the foragers who follow the shorter path make more journeys than the others and the trail they leave soon becomes the more scented and attractive.

What really is surprising is that, should a traffic jam arise, the ants can organize themselves to regulate the flow. This discovery, too, we owe to Deneubourg, working this time with Audrey

Dussutour from the Université Paul-Sabatier (Toulouse) on the black garden ant *Lasius niger.* They set up an experiment including a diamond-shaped bridge between the nest and the source of food. In this way they observed that, when both arms of the bridge were wide enough (ten millimetres) to allow for fluent traffic, the ants very soon came to prefer one of the two available paths. But when both arms were narrowed to six millimetres, the ants coming from the nest along the more frequented of the two kept colliding with those going back to the nest, who 'shifted' them over towards the other arm. In a very short time, the traffic was equalized on both arms. According to the authors of the study, 'The ants organize their traffic in a way which can be described as optimal', by avoiding jams and bottlenecks which might delay the arrival of the colony's supplies. This is how the Brussels team explain the ants' ability to take the shortest path back to the nest (see Figure 1).

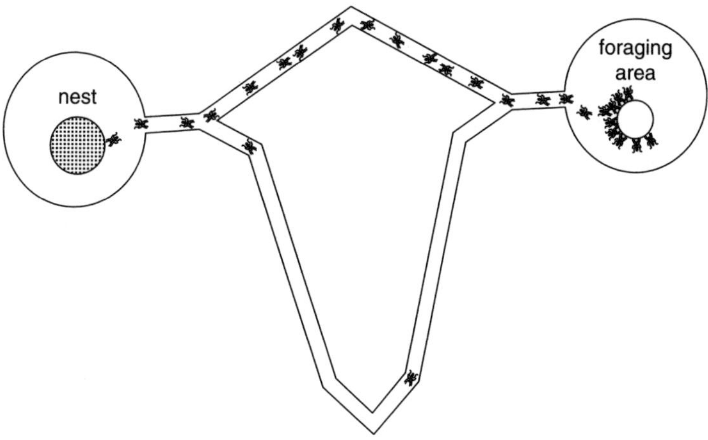

Figure 1 The shortest path An experiment devised by Jean-Louis Deneubourg and Serge Aron, from ULB, giving ants the choice between two routes to a source of food.

Several studies have now revealed that ants deposit a complex array of pheromone signals and that this allows them to provide information about more than just the path to follow. For example, in the Malaysian ponerine army ant *Leptogenys distinguenda*, temporal and spatial variation in the use of two distinct trail pheromones provides context-specific information in directing and organizing raids. A first pheromone produced in the poison gland is used for attraction to prey. It is highly volatile and lasts for only five minutes, ensuring that ants are not attracted long after a particular prey has been captured. In contrast, another trail pheromone, produced in the pygidial gland, lasts longer (approximately twenty-five minutes) and is responsible for maintaining the spatial organization of raiding ants. When attacking prey, workers often become detached from the trail network and this pheromone guides them back to the trail or the colony.

There are many circumstances in which it would also seem useful for workers to be able to deter foragers from taking unrewarding routes. To determine whether ants have such an ability, Elva Robinson, Francis Ratnieks, and colleagues from the University of Sheffield transferred paper substrates between trails leading to rewarding and unrewarding paths. These experiments, conducted on the pharaoh ant *Monomorium pharaonis*, revealed that workers use a negative repellent pheromone to mark unrewarding branches. The signal is concentrated at decision points (trail bifurcations). As it is volatile, it provides advance warning, like our traffic signs before road junctions. This is another example of the various sophisticated control mechanisms used in communication.

An alarm system

When activated by ants that have become aware of a danger threatening the colony, pheromones can also act as an alarm system which alerts nestmates. This can be easily demonstrated by disturbing wood ants: all workers present on the surface will instantly go into aggressive mode, squirting formic acid from their abdomens and becoming ready to bite any hand that might try to penetrate the nest. Some primitive ants also have recourse to this chemical language to call for help.

The communication systems of some species can actually be quite sophisticated. In the 1970s, the British entomologist John Bradshaw discovered that, when African weaver ants come upon an enemy, they can emit a mixture of four pheromones. These chemicals have different rates of diffusion in air; and they enable individuals to call for assistance and to guide reinforcements towards themselves. The alarm functions in stages: on perceiving the first of the pheromones released by the individual in distress, ants will begin waving their antennae to pick up other scents; they then receive the second chemical signal and start running about in all directions looking for the source of the problem, until they pick up the third signal, which attracts them in the right direction. When they reach the destination, they perceive the fourth of the pheromones, which has the effect of making them even more aggressive.

In addition, chemical language is a way for ants to mark their territory. Some, such as African weaver ants, use their own fecal matter, like cats and dogs. Unlike other ants, which 'defecate either in a remote corner of the nest or in a special garbage area outside the nest, a pile of detritus entomologists call the kitchen midden', the weaver ants defecate over the whole surface of their territory. When Hölldobler and Wilson gave a weaver ant colony access to an area that its workers had not visited before, 'the rate of defecation soared. At frequent intervals, far beyond what could have been their physiological need, the workers touched the tips of their abdomens (the extreme posterior ends of their body) to the surface and extruded large drops of brown fluid through their anuses'. When two colonies were confronted with each other, it was observed that the scouts' recruitment behaviour was much more intense on territory marked by the enemy than on territory impregnated with their own colony's smell.

Ant colonies thus take over territory which they are careful to mark with their own chemical imprint. If they encounter other species or colonies, they will certainly try to intimidate them; but these encounters seldom turn into pitched battles. The home range is not considered to be the colony's exclusive preserve, but rather 'an area that the colony knows to be hospitable and available for foraging,' as Cédric Devigne and Claire Detrain of ULB put it. So the *Lasius niger* ants refrain from marking it as their own, accepting the traces left by other colonies as evidence that this is a tract of territory rich in food sources. The term used by the Brussels entomologists for this behaviour is 'shared information strategy'.

Signs of recognition

Elementary observation of ant colonies going to and fro, some leaving the nest, some returning to it, shows that at every meeting between individuals they touch antennae, as though exchanging tokens of recognition—which is in fact what they are doing. Since all the inhabitants of a nest share the same smell and can thus easily identify one another, touching antennae is a way of telling who they are dealing with, of detecting the presence of a member of their own species or colony.

These smells come from chemicals that they spread on their cuticle. Wasps do this too, but the Formicidæ are alone in possessing a postpharyngeal gland (situated behind the pharynx), which acts as a reservoir of scent. Every ant produces a mix of pheromones of variable composition depending not only on its species but also on its individual genetic and physiological characteristics or even diet. Through physical contact among the ants, these individual smells blend into the collective odour of the whole society. It is this combination which becomes the sign of recognition for all members of a colony, enabling them to tell friend from foe.

There is more to it than that, though, for within a single colony, each caste has its particular olfactory profile. All members of a group secrete the same pheromones, but in slightly different proportions according to their stage of development or the caste they belong to. The queen has a quite specific smell, which differentiates her from her barren daughters; and every specialized caste has its own smell, as do eggs and larvae, to distinguish them from adults. For an ant to know who it is dealing with, all it has to do is touch a nestmate with its antennae, which will transmit distinct messages about the latter's social class. There is, however, no experimental evidence that ant workers can identify a single individual by smell. They appear to be incapable of telling Jane from Joan.

Dancing and squeaking

Though ants communicate largely through pheromones, they also have other ways of exchanging messages. When weaver ants, for example, want to take nestmates to a particular place outside the nest, they add to their chemical trail a set of gestures or acts including touching or even little dances, though these are much less sophisticated than the celebrated honey bee dance.

They can also give sound signals, made of a high-pitched squeak which they produce by rubbing a thin transverse scraper near the petiole (a constricted segment connecting the abdomen and thorax) against a washing board of fine parallel ridges on the adjacent surface of the abdomen. This stridulatory organ, as it is known in entomology, though not common to all species, is used for sending distress signals, particularly if an ant has been buried by the collapse of an underground passage.

For leaf-cutting ants, with the high consumption of plants required by the fungus they cultivate, this acoustic communication is also a good way of informing nestmates that they have discovered a good supply of leaves. It's a way of saying, 'Come over here! I've just found some very tasty ones, easy to cut.' We owe this discovery to Flavio Roces and colleagues, who offered laurel leaves of different quality to *Atta cephalotes* ants. When they were dealing with tough older leaves, only 40 per cent of the workers activated their stridulatory organ; 70 per cent used it with leaves that were finer and softer; and when they were given leaves treated with sugar, whether old or tender ones, all the ants, without exception, started their squeaking. Scientists wanted to verify that these acoustic signals did in fact attract ants that were chewing nearby leaves, so they continued their experiment by linking the diaphragms of two loudspeakers to two different leaf-bearing branches of laurel. They then played a recording of the stridulation first through one loudspeaker, then through the other: the branch linked to the one giving the signal was instantly crammed with ants; while the silent one attracted not a single ant.

Other species use these acoustical signals to reinforce the pheromones for recruiting nestmates when they have discovered large prey or a specially tasty leaf. According to Cesare Baroni-Urbani, Elmar Schiliger, and their colleagues at the University of Basel, 'Sound production through stridulation is an important but not essential component of ant recruitment behaviour.' In a study of ants of the genus *Messor*, the finding of this team was that any worker on whom they had fitted a silencer (by putting wax on its stridulatory organ) still managed to recruit as many nestmates as the others; it just took longer.

This stridulation is all but inaudible to the human ear, as its intensity is very low. The Swiss team report that they have never heard their *Messor* making the sound in the wild but that it is clearly audible if you take an ant by the head and hold it close to your ear.

Not that ants themselves actually 'hear' the sound; they pick up stridulatory vibrations through the ground in signals which only carry for a few centimetres. This is a clever way of summoning nearby workers without needlessly alerting the whole colony.

So, with their pheromones on the one hand and their stridulation on the other, ants can use signals which are effective over both long and short distances. This means they are well equipped to send and receive all the information essential to everyday life and the proper working of their colony. The constant exchanging of information is what has enabled colony members to exploit mutual assistance and cooperation to the full. As a rule, successful societies have good communication systems; ant society is no exception.

11
Family models

On the face of it, nothing could be simpler: ant colonies consist of a reproductive queen and all her daughters, and the workers, who are sterile. This makes for a neat straightforward model; and until about 1990, the majority of entomologists subscribed to it. Indeed, this explains why, for purposes of simplification, this book has been describing the social arrangements inside the nest in terms of 'the' queen and her daughters. But in reality things are much more complex than this. There certainly are colonies which live under the rule of a single queen ('monogynous' is the term used to define them); but there are some which have several queens ('polygynous'); and there others which have none at all. There are additional complications, in that some species have both monogynous and polygynous colonies.

Unmarried mothers

There are about a hundred ant species which do not have queens. In such species, all the female workers have the ability to mate and reproduce, in marked contrast with species that have queens, in which workers are morphologically incapable of mating. It seems that these queenless ants derive from species that once had both queens and workers. However, because they lived in habitats that were dry and inhospitable, it was difficult for young virgin queens to fly off to find a suitable environment in which to found a new colony. Through evolution, this mode of reproduction by queen dispersal presumably died out and the morphological differences between the sovereigns and their daughters were gradually modified until only workers were left. Nowadays such colonies contain only workers, but these have kept a spermatheca and thus are in principle all capable of breeding. However, this ostensible egalitarianism does not mean that every female manages to reproduce. On the contrary, the denizens of a single nest, as a way of establishing their dominance, engage in fights which may be lengthy and fierce, after the manner of many mammals. The difference is that, among mammals, the fighting is mainly done by the males. Among these ants, the only workers to mate and reproduce are the winners.

Even so, queenless ants are still rigidly hierarchical. This can lead to some surprising behaviour, such as that seen among *Dinoponera quadriceps* or the South African *Streblognathus peetersi*. At the end of their fighting for dominance, as Christian Peeters and Thibaud Monnin of the CNRS in Paris have observed, a sort of 'ranking' is established. The winning worker is 'crowned' queen, or rather, 'gamergate' (from the Greek *gam*, 'married', and *erg*, 'work'), as Peeters prefers to call these 'married workers', to avoid confusion with the 'true' queens. The fighting then starts again, to decide the runner-up, third place-getter, and so forth.

Once the winner is clearly accepted, its ovarian activity sets in. It acquires 'not only a layer's physiology but also a fertility pheromone', according to Virginie Cuvillier, an expert on *Streblognathus peetersi*. This pheromone, which is on the cuticle surface, provides information on social status. As long as the gamergate is covered with enough of the fertility pheromones, she is protected by the workers of lower rank. But if not, she must beware, as entomologists have shown through the use of hormonal treatments which decrease the fecundity of the dominant. Very quickly the runner-up noticed the change and started to become aggressive. Within a few days, her take-over had succeeded: the lower ranked workers had deposed the deficient layer, replacing her with her closest rival. 'This is why some have described this type of social organization as a meritocracy,' says Christian Peeters.

Such competitiveness is useful, in that if the gamergate should die, the runner-up takes her place and starts reproducing. Even at this stage, fierce fighting can recur, as the lower place-getters will pick a fight with the new sovereign whom they see as a usurper. In such circumstances, other members of the colony step in at once to settle things down; several of them will hold each of the combatants by the legs until they are calm and law and order are restored to the community.

Other species have other methods. Among primitive Ponerine *Diacamma* ants from Australia, Japan, India, and Malaysia, all individuals are born with a pair of innervated thoracic appendages termed gemmae, which are in fact glands for the production of domination pheromones. To prevent any possibility of competition, the dominant female systematically mutilates the gemmae of the newly emerged workers, which then will never be able to become dominant and reproduce. When she dies, she will be replaced by the first unmutilated newly emerged worker, who in exactly the same way sets about mutilating all the others who emerge after her. Then the cycle will start all over again.

Several million daughters

All that said, the simplest family structures are still those of monogynous ants, in which a single queen reigns over a numberless profusion of daughters: colonies of army ants, for instance, can contain, at any given time, several million workers. What this means, of course, is that over her lifetime a mother ant may give birth to several million daughters.

However, quite a few species which were once thought to be monogynous have turned out to be polygynous. A case in point is *Pheidole*, whose underground nests are so difficult to excavate that it can be even more difficult to find their well hidden and protected queens. Working with Denis Fournier and Serge Aron from ULB and Frédéric Tripet from the University of California (Los Angeles), we managed to uncover their secret family structure. By analysing the workers' genome, we were able to identify the genotype of their mothers and thereby conclude that there were at least two species of the genus *Pheidole* which were capable of forming colonies with two to four queens, all coexisting under the same roof.

Though only a few species of *Pheidole* (at least in the present state of knowledge) live polygynous lives, there are other groups in which polygyny is more or less the norm. For example, among *Myrmica* ants, well known for their painful stings, cohabitation of several queens is common. It is estimated that between 50 and 60 per cent of them have adopted what is called 'facultative polygyny'. The number of queens, which can be anything between one and ten in the same nest, varies according to species and also between colonies within species.

Similarly, wood ants form polygynous colonies; for instance, more than 95 per cent of *Formica exsecta* colonies contain multiple queens, as we established by studying them in the wild. One spring, when the insects came to the surface of the nest, we captured several queens, which we marked by tying a thread of extremely fine metal between their thorax and their abdomen, a most delicate operation requiring a fair amount of neat fingerwork. Having released them, we did the same thing week after week, a technique of marking and recapturing through which we discovered that, on average, there were about 100 queens per colony. The record is held by another species of wood ant from the Jura uplands, which was found to have more than 1,000 queens in a single nest. There does seem to be a correlation between the proportion of polygynous nests and the average number of queens per nest. In other words, in a species with mainly polygynous nests, you often find nests containing a great many queens.

As for the Argentine ants now established in Europe, the United States, and several other parts of the world, they too have many queens in their colonies. There is however, constant movement of individuals between nests, the boundaries of which are not clearly defined. 'Unicolonial' is the word used to describe this type of social organization, in which it is difficult to define accurately what constitutes a nest, let alone try to count how many queens there might be in each one.

The family structure adopted by ants depends in large measure on the environment, polygynous societies being more common in colder or temperate regions than in tropical areas. There is a straightforward explanation for this: ants living in colder climates have a harder life than those living in the tropics; food is not as easy to find and they must often move their nests, which is not without danger for a queen. Virgin queens, too, run a greater risk of meeting accidental death when they fly away from the nest; and any who survive have greater trouble founding their new family. As a way of limiting these risks, it is in the interest of a society to develop by 'budding', that is, by welcoming members of other colonies, including their queen.

Some species of small *Leptothorax* ants, which live in dead twigs in coldish parts of North America, are a good illustration of how environment can influence the way societies are organized. Colonies established in vast homogeneous forests, where there is little risk of lacking shelter or food, often have a single queen. But those established in places where growth of trees is not as uniform show a tendency towards polygyny. This can explain why closely related species, or even colonies belonging to the same species, may have adopted different ways of life. During the course of evolution, the general trend is a shift from monogyny to polygyny, from absolute monarchy to power-sharing.

Of queens large and small

The number of queens living in a single nest also has a bearing upon the way in which a colony is founded; and that can have a corresponding effect on the size of the queens. In the world of ants, a mother is always larger than her daughters. But this difference is more marked in monogynous species than in polygynous ones. There are, for one thing, simple practical reasons for this. If a queen is the sole reproducer of the whole colony, she must lay a greater number of eggs, which means she must have larger ovaries. A second thing is that, in most monogynous species, the queen is charged with the heavy responsibility of starting a new society without the help of workers. This means she must cope all by herself for several months, the time it will take for her first-born daughters to reach adulthood and become proper workers. Sometimes, as in the case of primitive ponerine ants, she hunts prey so as to feed her first generation of descendants. But in most species she stays in the nest, which means she has to have already accumulated enough energy before mating flight especially in the form of lipids, to be up to the task. Polygynous colonies, on the other hand, develop quite differently. Young queens often return to an established colony where workers will immediately help them to raise their brood. These young queens lead an easy life and so have no need to accumulate fat before the mating flight.

In some species, the cohabitation poses no apparent problem and the queens seem to get on well together, as happens among wood ants. With the technique of marking and recapturing that we used for studying the colonies in the Jura, we observed that the great majority of the queens reproduce and that, with slight differences, they have about the same number of offspring.

But it can also happen that some representatives of the aristocracy brook no rivals when it comes to reproduction, and so go in for violent fighting. This was demonstrated in the 1980s with a species of tiny *Leptothorax* ants of North America. Though their queens command small populations of only fifty to 100 individuals, they have fierce set-tos at the end of the winter. The dominant female will monopolize the bulk of the reproduction, while the others will either remain in the nest, eking out a subordinate life with all the characteristic behaviour of plain workers, or else will be simply evicted from the nest. It must be supposed that this amounts to a death sentence.

Some queens outwit the others. Without ever fighting, they increase their own reproductive success by specializing in the production of new queens instead of workers. A study of the red imported fire ant by Kenneth Ross from the University of Georgia (USA) made an amazing finding: not long before they die, the mothers tend to produce a greater number of baby queens. In this way, they manage to bias the process determining the destiny of the larvae, which is usually left to the workers. How do they do this? Why do they employ this subterfuge in their dying days? Is it age that helps them to accomplish this trick? Or is it just that the production of future queens wears them out and hastens their demise? We have no answers to these questions; as in so many other areas, this behaviour is a mystery to myrmecology.

12
Parasites and slave-makers

That ants are social creatures cannot be doubted. However, some of them seem to have forgotten the proprieties of social life. Their queens, instead of founding their own families and living amid their daughters, set up house in an alien colony where they behave like squatters. They live there as parasites and can even behave like slave-owners.

Ants of the genera *Formica*, and to a much lesser extent *Lasius*, frequently adopt behaviour known to entomologists as 'temporary social parasitism'. In these species, which are often polygynous, the queen is small and does not accumulate enough fat to be able to found a colony by herself—not that she needs to, for she just enters an established nest of a closely related species. There she starts by duping the workers, which she does by covering herself in their smell; then she kills the resident queen, takes her place, settles down and quietly sets about laying her eggs.

This parasitism is called temporary because once the new queen is well established in the purloined nest, she produces her own family, which gradually replaces the whole colony. The ownership of the nest having changed, its new occupants lose their status as parasites; and so after a while, everything in the nest reverts to normality. In about eighty species, however, the parasitism lives on, in what is known as 'permanent social parasitism'. These queens, generally small in size and often without wings, also enter the nests of other species. But once inside, they do not kill the resident queen; they just exploit the resident workers, which raises their brood for them, consisting essentially of males and future queens.

An extreme example of such behaviour has been observed in *Teleutomyrmex schneideri*. These ants are few and far between, having been located in only four areas, all in high country: in the Swiss and French Alps; on the Pic de Fabrège in the Spanish Pyrenees; and in Turkmenistan. This is a species which is completely without workers and in which the morphology of the queens is very peculiar, in that they have a concave abdomen which enables them to settle on the abdomen or the thorax of the host queen and live out their whole life there. Living in total dependence on their host species, *Teleutomyrmex schneideri* have small brains and their metapleural gland has degenerated, as has their sting. Though they are queens, these ants are no larger than the workers of the host species whom they resemble, a form of mimicry possibly allowing them to fool the nursemaids and pass unnoticed.

Brood theft

Other forms of social parasitism can reach the point of virtual enslavement, the simplest of which is known as facultative enslavement. The origins of a colony follow the traditional model: a queen leaves a nest, founds her own society unaided, and produces her workers. The latter, however, instead of being content to tend their own brood, are quite prepared to resort to kidnapping. From time to time, they enter a nest belonging to ants of their own species, or even of a different one, and make off with the larvae and the pupae. This behaviour is especially marked in *Formica sanguinea* slave-making ants, known as blood-red ants and also in *Myrmecocystus mimicus* (the honeypot ant). Genetic analyses done by Daniel Kronauer have established that the workers in a third of the *Myrmecocystus* colonies were not biological offspring of the queen but slave workers kidnapped from other nests.

By contrast, there are fifty-five known species, belonging to the sub-families Formicinae and Myrmicinae, in which enslavement is in no way optional but utterly unavoidable. In these cases, it is the queen who initiates the parasitism after the mating flight by entering, unaccompanied, the nest of an alien colony. Howard Topoff and his colleagues from the Museum of Natural History of Arizona actually built a transparent nest in the laboratory, which enabled them to give a detailed description of the arrival of a young queen of *Polyergus breviceps* in a colony of *Formica gnava*. Their report on what ensued reads rather like the account of a successful bank hold-up:

> 'In most cases the *Polyergus* queen quickly detects the entrance and erupts into a frenzy of ruthless activity. She bolts straight for the *Formica* queen, literally pushing aside any *Formica* workers that attempt to grab and bite her, . . . using her powerful mandibles for biting her attackers and a repellent pheromone secreted from the Dufour's gland in her abdomen. With the workers' opposition liquidated, the *Polyergus* queen grabs the *Formica* queen and bites her head, thorax and abdomen for an unrelenting twenty-five minutes. Between bouts of biting she uses her extruded tongue to lick the wounded parts of her dying victim. Within seconds of the host queen's death, the nest undergoes a most remarkable transformation. The *Formica* workers behave as if sedated. They calmly approach the *Polyergus* queen and start grooming her—just as they did their own queen. The *Polyergus* queen, in turn, assembles the scattered *Formica* pupae into a neat pile and stands triumphally on top of it. At this point, colony takeover is a done deal.'

Once again the usurper, by licking the resident queen and acquiring her pheromones and thus her smell, contrives to be accepted by the workers. The Arizona team also observed that the *Formica* workers will not accept the newcomer if their queen is absent from the nest. Instead they attack the *Polyergus* with their mandibles, pinion her by the legs, and bite her to death.

Once she has usurped the throne, the intruder feels at home and starts laying. Over the course of evolution, however, her worker daughters have lost the ability to care for the brood or even to find food. To begin with, the *Formica* do these tasks for them; but then, as they gradually die out, the colony is left without labourers. To obtain new slaves to do the work required by the colony, the *Polyergus* workers turn to a life of kidnapping and fetch in larvae and pupae from neighbouring colonies of *Formica*. Even the morphology of the *Polyergus* criminals has adapted to their lifestyle: they have strong cuticles; and with their curved sabre-like jaws they can cut through the heads of any ants that might want to object to their kidnappings. A single nest of such *Polyergus* Amazons, which may contain 2,500 ants of that species, may also contain up to 6,000 slaves. The kidnappers certainly know how to conduct their business: they all leave the nest together and gangs of hundreds of them will cover great distances trying to find some other nest. When they find one, they recruit huge numbers of nestmates, forming battalions of 600 or even 1,000 individuals, which then launch an invasion.

Inside the targeted nest, the arrival of these plunder squads generally leads to little fighting, because the slave-makers use a range of ruses and tricks to get in. Some of them use the standard technique of taking on the smell of the nest they are attacking, which they do by grabbing a worker and smearing their bodies with the exudation from its cuticle. Some, such as *Polyergus topoff*, produce pheromones with a calming effect. Others act by secreting what some entomologists call 'propaganda substances', molecules that function like false alarms, making the residents of the nest flee in disarray. Then there are *Harpagoxenus sublaevis*, whose specialty is the total demoralization of the colony they intend to take over. When they enter a nest of *Leptothorax*, they do not fight the resident ants. With their stings they dab a substance on the bodies of the nest-owners which makes them turn violently against each other and fight to the death among themselves. The ensuing free-for-all gives the intruders the chance to snatch the brood and restock their own nest.

These slave-maker species are in general very selective in their raids; they do not make indiscriminate attacks, having favoured over evolutionary time a mode of specialization that restricts their targets. If, say, queens of the European species *Polyergus rufescens* enter a nest of *Formica rufibarbis*, the attempt is doomed to failure and they are killed by the resident workers. But when they attack *Formica cunicularia*, the success rate is very high: they achieve their aim in 85 per cent of cases. The workers, too, are selective in their house-breaking and will not attack just any species. When they do join in, they wreak havoc in the colonies of their victims: in the late nineteenth century, the Swiss entomologist Auguste Forel observed that, in one season, a single *Polyergus* could steal up to 40,000 larvae or cocoons in a *Formica* nest. Recently, a genetic study also revealed that about half the nests of *Temnothorax longispinosus* were being burgled by *Protomognathus americanus*. In the nests of the latter a mixing of species takes place which can lead to cases of co-evolution.

Genetic similarity

Parasitic and slave-making ants are usually genetically close to the species in whose colonies they squat or whom they plunder.

This closeness goes by the name of 'Emery's rule', from the name of the Swiss-Italian entomologist Carlo Emery who first defined it in 1909. Taken strictly, the rule means that both the attackers and the victims are next-door neighbours in the tree of evolution. It suggests, too, that the host ants and the intruders must originally have belonged to the same species. At that stage, their colonies, which were polygynous, recruited some new queens after they had mated. Some of these young queens must eventually have become real parasites, producing mostly or solely reproductive individuals (queens and males). At a later stage during evolution, some descendants of these queens would then have mated only with their own brothers; and it was this 'reproductive isolation' which gave rise to a species that was different from the original one.

Genetic studies have shown that, in some cases, the strict form of Emery's rule does indeed apply. However, in others, the parasites and their victims, though related, belong in fact to more distant family branches, which means they conform to a more relaxed version of the rule. How such cases arose is not known. It may be that some queens, having evolved into parasites, became attached only to species evolutionarily distant from their own. Or it could be that, to begin with, they plundered nests of closely related species, only to change tactics (and host species) during evolution.

However it came about, these parasitical and kidnapping types of ants are much more widespread in northern latitudes than in the tropics. Parasitism and slave-making tend to develop in the most hostile ecosystems, where young queens have the greatest difficulty in founding families of their own. This is why they seek to exploit the resources of established colonies, which can be seen as further confirmation of the fact that ants are particularly adept at adapting to their environment.

Part III

Nowt So Rum as Ants!

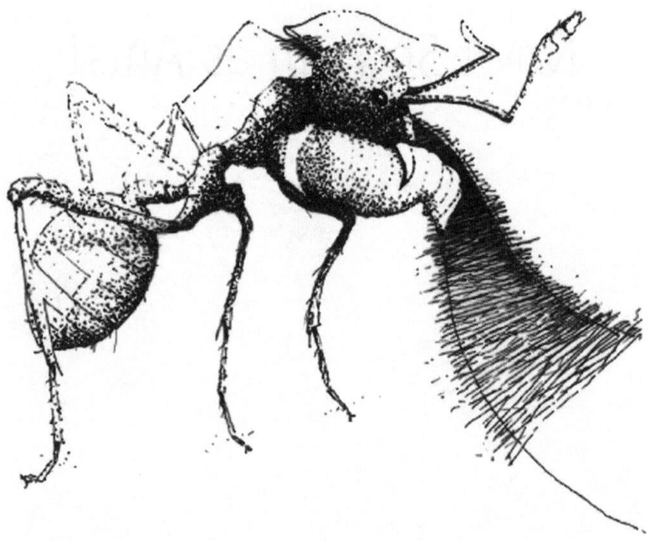

Drawing 3 Weaver ants (Oecophylla) Weaver ants live in the tree-tops, where they build nests out of layers of leaves. To glue the leaves together, they use the silk secreted by their larvae, which they move back and forth like the shuttles of looms.

13
Army ants

Every society has its outstanding personalities, its stars, who swagger through life and are made much of. The galaxies of the ants are no exception, for they too have their stars, extraordinary luminaries with original ways of doing things and antics that prove very attractive to myrmecologists. Scientists study these species very closely, at times with astonishment and admiration at the devious ways in which some of them have contrived to adapt to their milieu, and sometimes with anxiety at the ravages they cause. Weaver ants, for instance, can astound even the experienced entomologist with the skill they show in stitching leaves together to make their nests high up in the canopy. Honeypot ants, too, can be the source of much amusement as they gorge themselves on sugar and act as the colony's larder.

Rampaging nomads

Or take army ants, charging about in dense phalanxes, living off the land, striking camp every day like hardened infantrymen. These American and African species are known as army ants because when they set off on the hunt for food, they do so in great numbers, marching onwards like a real army, wreaking havoc among any insects and arthropods that happen to live in their path. William Molton Wheeler called them 'the Huns and Tartars of the insect world'. And it is a fact that army ants are particularly formidable, being among the main predators in the tropics. Even more remarkable is the ingenuity they display in their life history and in their ways of feeding and reproducing, which set them apart from all other ants. These behaviours explain why there is what is known in entomology as 'the army ant syndrome'.

These marauders would appear to have derived such behaviours from a common ancestor that lived about 110 million years ago, according to Sean G. Brady of the National Museum of Natural History, Smithsonian Institution, Washington, not long before the two continents of Africa and the Americas split apart. After the separation, they evolved into three different sub-families. The first of these, Ecitoninae, which is found in America, contains about 150 species, including those in the much-studied genus *Eciton*. The other two, Aenictinae and Dorylinae, account for about 100 species and are the African branch of the family.

Whichever side of the Atlantic they live on, all army ants have the same lifestyle, which is nomadic. They alternate periods of travelling, which can last up to a fortnight or three weeks and during which they move camp every night, with sedentary periods of longer or shorter duration, depending on species. This rhythm of life leaves little scope for building the sorts of nests that other ants build. In fact, it is one of the specific features of army ants that they have, strictly speaking, no nest. They live in what have been called 'bivouacs', a term coined by Theodore Schneirla of the American Museum of Natural History in New York City, one of the first entomologists to make a proper study of them. With its echo of the military life, the term 'bivouac' is certainly a neat description of the temporariness of their accommodation. However, the comparison with human armies goes no further, for the bivouac is not just thrown up by the workers, it is actually made of workers. By clinging to each other, they form a dense mass which, in the *Eciton* of Costa Rica, for example, is cylindrical or elliptical in shape and one metre in diameter. There may be half a million individuals in this structure, inside which the queen and her brood are sheltered.

Instead of camping out, the African *Dorylus* prefer to be housed underground. Some of these species, such as *Dorylus nigricans*, which may remain sedentary for four months between two migrations, do actually dig out a sort of nest, though it is pretty rudimentary. The workers may excavate a few galleries; but they usually adapt to natural cavities in the ground, which they merely extend into chambers. At times, they will even reoccupy a nest that they or some other colony of their own species have already lived in, makeshift housing that they leave every morning to go on the hunt.

Committed carnivores

With the sole exception of *Dorylus orientalis*, who are herbivores renowned for the devastation they can cause in crops, all army ants are 'uncompromisingly carnivorous', according to William H. Gotwald, Jr, in his book *Army Ants: the Biology of Social Predation*. Their favourite prey are, 'in descending order of importance, ants, termites, and wasps. This bill of fare is generously supplemented with a wide variety of other invertebrates and occasional vertebrates'. Spiders, scorpions, cockroaches, beetles, grasshoppers, and other arthropods can expect no mercy; nor indeed can other much larger creatures. In 1959, a Jesuit, Father Albert Raignier, reported that in the course of a single night a column of ants had consumed about ten hens, five or six rabbits, and a sheep. It is said that in Brazzaville *Dorylus* ants once ate a lion in a cage, leaving only its skeleton. This is no doubt something of a tall story; but it is beyond doubt that any animal that is injured, hobbled, or enclosed may become prey for these all-consuming marauders of the tropical forest.

It must be said that their hunting technique is particularly highly developed. In most ant species, the search for food is led by a few scouts and the nestmates are not alerted until a source of it is found. Army ants set about things very differently. They always seek their prey in groups; and it is during these expeditions that their cooperative spirit is seen at its best. At daybreak, the living sphere of the bivouac starts to disintegrate, breaking into teeming chains and clumps of workers that drop to the ground. At a marching pace of twenty metres an hour, they leave their campsite in a long column which eventually splits into several strands. There is also the 'swarm raid', which entails leaving the bivouac in different squads then forming up into a fan-shaped body of ants. Schneirla made many observations of *Eciton burchelli*, one of whose expeditions he describes as follows: '[They form] a rectangular body of fifteen metres or more in width and one or two metres deep, made of many tens of thousands of scurrying reddish-black individuals, which as a mass manages to move broadside ahead in a fairly direct path.'

The hunters stop work about noon and return to the bivouac. And some work it is, too! According to the British myrmecologist Nigel Franks, from one of their raids, which may take them up to 100 metres away from their starting point, *Eciton burchelli* bring back, on average, 30,000 prey. In this species, workers are extremely well organized and very skilled in task-sharing. They are divided into four castes, morphologically very different from each other, and all with specialized functions. The smaller castes, the minors and the media, whose normal job it is to look after the queen and the brood, also play a part at hunting time. They run along the chemical trail, reinforcing it with deposits of pheromones, and capture prey. The larger workers, the majors, are not as nimble; but with their long strong mandibles, they make perfect soldiers when the colony needs to be defended. The submajors, who are not quite as big as the majors, have the onerous responsibility of transporting the prey. Though they make up a mere 3 per cent of the worker population, Franks calculates that they account for 26 per cent of the bearers: 'Submajors have the longest legs in proportion to their body size of any *Eciton burchelli* workers and they can run faster than any of their sisters and can carry disproportionately large items.' Not all species of army ants take division of labour to the lengths seen in *Eciton burchelli*. But all, given the sheer numbers of ants in their colonies (ranging from several thousand individuals to several million in the largest), given their appetite, and the efficiency with which they carry out their raids, can have devastating effects wherever they go.

Customers, guests, and associates

These massed attacks, which few prey can survive, can be of benefit to other species, especially those which prey on the army ants themselves. Snakes, lizards, mongoose, gorilla, chimpanzees, and even other species of ant, can all find plenty to eat when they come upon a battalion on the march.

Nor are these the only ones to take advantage of an army of ants on the rampage, whose movements are, as Gotwald says, accompanied by 'a veritable menagerie of arthropods and vertebrates'. Though these animals neither attack the ants nor help them in any way, they still benefit from their presence. There are types of mites (a sub-group of arachnids), for instance, that hitch a lift on top of the advancing army and manage to make good ground through the forest or across the savannah. Other camp followers just take the opportunity for a free meal, like certain types of birds which fly close to the ground and can take an abundance of insects dislodged by the ants. The birds are followed by butterflies, which feed on their droppings. Then there are swarms of parasitical flies, which lay their eggs on the fleeing prey, and many others. Every colony of army ants is accompanied by thousands of these free-loaders, or 'guests and associates' as Gotwald calls them, in a great congress of species and genera.

With an army of such proportions, and with so many free-lunchers, the ants make short work of whatever feed is available in the vicinity of their bivouac. It is not only hunger, however, that makes them move on so regularly. In the 1930s, Schneirla observed that they migrated even when there was still prey to be had. In fact, the alternation of sedentary periods with nomadic excursions is closely related to a colony's reproductive cycle, at least in some species such as *Eciton*.

The static period, which in *Eciton hamatum* lasts for eighteen to twenty-one days, is a time of relative calm. The daily round of breaking camp and moving on is put aside; and though the ants still make sorties, they throw less energy and fewer workers into them. Within the peace and quiet of the bivouac, the queen can devote all her time to laying her eggs, which ensures the supply of new larvae. It is also during the sedentary period that the preceding generation of larvae will reach maturity and the adult workers will emerge from their cocoons. This development gives a new surge of energy to the whole colony; there is an increase of activity in and around the campsite; the forays start to last longer and become more intense. Then, with the regularity of a metronome, the cycle takes up again as the colony goes back into nomadic mode.

Eggs by the million

The amazing originality of army ants is apparent in more ways than one. The queen, for example, 'is beyond doubt the most atypical queen of all ant species', as Gotwald puts it. She is 'so unusual-looking that even an experienced entomologist might not recognize her as an ant'. These queens are completely or practically blind; they are not only much larger than the workers, but are quite unlike them in appearance too. Their rate of laying is also phenomenal: a single queen of *Eciton burchelli* can produce 2.4 million eggs per year, an astounding performance, yet one which pales into mediocrity when compared with *Dorylus (Anomma) wilverthi*, the all-time record breaker at three to four million eggs during the sedentary phases, which makes a total of several tens of millions per year.

Another oddity is that the queens have no wings. They have no need of them, as they never leave the nest for a mating flight. Males seek them out, courting them and seeking their favours in great numbers. Whereas most queens of most ants mate only once or a few times, queens of *Dorylus (Anomma) molestus* may mate with twenty different partners.

This peculiarity of the queens is matched by that of the males. They are well-built individuals, much larger than workers, with 'uncommonly well-developed genitalia', says Gotwald, and with mandibles 'that are often formidable in appearance, although the males seem unable to inflict a painful bite'. He adds: 'I have frequently collected males with impunity and have never been bitten.' Some males fly away soon after emerging from their cocoons; others stay a little longer in their families of birth, but they too will abscond at the onset of the migration phase, for each and every male is destined to seek out a foreign colony. On arrival at such a colony, they get rid of their wings and join a column of hunters, in an attempt to be adopted by them. As things turn out, it is always the workers who, by accepting this or that male rather than some other, select the father of their future half-sisters. The fortunate bridegroom can then enter the bivouac, where he will stay for some days or weeks before mating with the queen in circumstances which are 'unclear', according to entomologists, who have as yet not managed to observe how it is done. After this, the male dies.

Once mated, the queens do not attempt to found a new family on their own. Left to themselves, they could never supply the needs of their society, since building the bivouac and hunting in large parties require a queen to be constantly surrounded by a great many workers. So new colonies are created by splitting: the population of workers divides into two roughly equal parts, one of which will stay with the queen mother while the other one will go with the young newly mated queen to establish a new colony. The two new families part company and live henceforth as strangers to one another. In this moment of choice, it is the workers who decide which of the queens, their mother or their sister, will command their future allegiance. So it is that, just as they select the individual who will father their siblings-to-be, they also select the one who will produce their unborn companions in wandering and marauding.

14
We work at the weaver's trade

Like army ants, weaver ants live in huge colonies which may number up to half a million individuals. Another similarity is that weaver ants have a very marked sense of cooperation. However, whereas army ants turn group cohesion to great advantage in their hunting techniques, collaboration in weaver ants is at its best when they are constructing their nests up in the treetops. In fact, their name derives from their building technique. The nest starts very simply. A group of worker ants finds a leaf that is soft and easy to bend. Several ants then line up, each holding an edge of a leaf in its mandibles and feet. Slowly they pull two leaves together. More and more workers join in, linking their feet and pulling until the edges of the two leaves are nearly touching. Other workers now carry larvae from the old nest and gently squeeze them between their mandibles, causing the larvae to ooze a thin thread of silk. Then the workers get busy, stitching the leaves together, just like tiny tailors. In fact, another name for weaver ants is tailor ants. All this makes for labour that is both Herculean and very intricate, requiring a high degree of collaboration and an elaborate system of

communication. It is a way of doing things which has no parallel among other ants or for that matter anywhere else in the animal kingdom.

When it comes to weaving, the out-and-out champions are the species of the genus *Oecophylla*. These are green or brown, quite large (the biggest of them can reach eight millimetres in length), and they are the most abundant tropical insects outside the Americas. A single species, *O. longinoda*, has managed to colonize most of the forests of Africa; and another one, *O. smaragdina*, close to *O. longinoda* on the tree of evolution, has extended its territory from India to the Solomon Islands and Australia. The spread of the genus *Oecophylla* over the surface of the earth is not a recent phenomenon. For example, two species have been discovered fossilized in amber from the Baltic, showing that they lived thirty million years ago during the Oligocene. Another fifteen million year-old fossil, of an entire colony, has come to light in Kenya. Analysis revealed that there was already a system of three castes, majors, media, and minors, very close to what can be seen in extant species.

It is clear that weaver ants have had plenty of time to prosper and colonize the warmer parts of the world. Their ecological success can also be explained by the fact that they have no shortage of space in which to found and expand their colonies—*Oecophylla* make their nests on the tops of trees and among their branches, which means their territory can stretch over a broad area of the canopy. Bert Hölldobler, in fieldwork in Kenya, has observed a colony which occupied no fewer than seventeen large trees. On a human scale of comparable complexity of organization, he says, 'the weaver-ant hegemony would be the equivalent of control by a mother and her children of at least 100 square kilometres of terrain'. The operative words here are 'at least', since the area occupied by the ants is not limited to the corresponding area on the ground: it also includes all the vegetation from the treetops to the ground, down to the last millimetre of leaf, branch, and trunk.

So weaver ants make their home in the trees, establishing their nests at the very top, for which they use remarkably original building techniques. Just imagine the initial choices that have to be made in selecting the aptest spot or the most suitable building material. A few isolated workers explore promising looking areas in the colony's territory, tugging at the edges and points of leaves, testing them for consistency and flexibility. If one of the ants manages to fold a leaf or to line up the edges of two leaves, her sisters working nearby instantly come to lend her a hand (or a leg, actually). They all link up like a tug-o'-war team and pull together to join the edges of the two leaves. If it turns out that the leaves are too far apart for a single individual to grasp them, they just form a living chain, or a bridge to be more exact: gripping each other by the petiole, as though holding one another by the waist, they bridge the gap between the two leaves. If one chain is inadequate to the task, they form several chains which work side by side, a veritable fabric of insects busying themselves in the treetops. The manoeuvres are delicate, requiring all the workers to pull at the same moment in the same direction, and this requires perfect coordination among the members of the building team.

Very super glue

All this is amazing enough, but the weaver ants have further surprises in store, especially as concerns their neat way of making sure their arboreal dwelling is a solid one. Bringing leaves together and joining them up is one thing; but it is quite another to make them stay together, since the merest puff of wind could blow the whole thing apart. These clever constructors have found the right glue for the job. The first to have observed this was Sir Joseph Banks, a member of Captain Cook's expedition to the South Seas in 1768. In Australia, Banks, who had a fine eye for such things, was intrigued by the way *Oecophylla* built their nests and noted in his journal that some workers 'drew down the leaves', while 'others within were employed to fasten the glue'. The real nature of this glue was not discovered till 1905, when the German zoologist Franz Doflein managed to observe that it was in fact the silk secreted by the larvae. The workers, generally majors, take the baby ants carefully in their mandibles and manipulate them much as a weaver does his shuttle. They move them back and forth along the edges of the leaves. In an attempt to observe the weaving process more closely, Hölldobler and Wilson filmed and analysed the complete sequence, frame by frame. What really surprised them was 'the rigidity with which the larva holds its body', turning itself into 'a largely passive instrument of the adult worker that has borne it from the interior of the nest'. From time to time, the larva stretches its head as it comes near the surface of the leaf, 'but otherwise it stays immobile and merely spins silk'.

By so doing, the larvae deprive themselves, for the sake of the colony, of the protective thread from which they make their cocoon. They are compensated for this by the shelter afforded by the nest where, even though uncocooned, they will manage to develop into queens, males, or workers. In the whole ant world, this is one of the rare cases of males taking some part in the workaday life of the colony, their role being normally restricted to that of reproducing. But even in this case the males are less cooperative than their sisters, for they have been shown to have smaller silk glands and contribute substantially less silk to the making of the nest than the female larvae.

Oecophylla are not the only ants to construct nests from leaves or to use their larvae as tools. Other weaver ants, such as *Polyrhachis* in Australia, do the same, albeit with techniques of lesser sophistication. They do not move the leaves around, but link them with walls of silk and various kinds of vegetable debris. There are also *Camponotus senex*, from humid forested regions of Latin America: they build dwellings containing networks of chambers and passageways but with the outside and inside walls made of silk, leaves serving as material only during construction. The techniques used by Brazilian *Dendromyrmex* ants are even simpler: they just keep reinforcing the leaves of their shelters with a layer of silk which is laid down by the larvae sometimes without the intervention of the workers. Nor has any entomologist ever observed *Dendromyrmex*, *Camponotus*, or *Polyrhachis* forming a living chain after the manner of *Oecophylla*, who have taken the art of nest-building in the trees to new heights.

Oecophylla do not restrict themselves to making mere huts out of leaves. They also build into the structure a great many rooms and tunnels, all of them woven with silk. Nor do they make do with a single hut, but spread a veritable network of nests across the tops of the trees. The queen's quarters are located near the roof of the canopy. The living quarters farthest from the centre of the colony are reserved for the oldest workers. It is their job to be on the watch for any intruders who might trespass onto the colony's territory, to attack and repel them. Weaver ants are in fact jealous keepers of their part of the forest and are fierce in its defence. The only outsiders welcome in the nest are insects such as mealy bugs, which the ants do more than tolerate: they actually protect them, as they find their excreta very much to their taste.

As can be imagined, a social organization of such complexity requires an efficient system of communication. *Oecophylla* have no fewer than five different ways of recruitment, whether for defending their society, reconnoitring new territories, harvesting food supplies, or of course making the nest. As in other species of ants, exchanges of information are effected through the chemical signals of pheromones, though *Oecophylla* also rely on straightforward touching of each other. A good example of this can be seen in the 'dialogue' between workers and the shuttle-larvae. The worker first feels the surface of the leaf with the tips of her antennae, before touching the larva's head to the leaf. Then she sets her antennae vibrating about the larva, tapping its head several times to make it secrete its silk.

Pheromones are the vehicle of communication between the different dwellings making up the network. The queen in particular signals her presence with a 'royal pheromone' which some of her daughters transmit to others, by touch, as they move about among the various nests. In this way, though their different living quarters may be several metres apart and even if they have no direct contact with her, all the workers are aware that their mother is present in the nest. The cohesion of the colony is thus assured.

15
Navigators who never lose their way

On the high plains of Afghanistan, the mounds made by *Cataglyphis* ants are covered in little stones, among which one can sometimes see specks of gold. This may well be the origin of the legend, recorded by the Greek historian Herodotus and later taken up by the Roman naturalist Pliny the Elder, that these ants are 'gold diggers'. Sad to say, there is no truth in this. In modern entomology, *Cataglyphis* go by the more prosaic name of 'desert ants', for the simple reason that they live in arid regions. However, they could just as well be called 'navigators', for they have an innate sense of direction. Like good sailors, their eyes even have what amounts to a compass.

Your standard ant finds its way back home using the Tom Thumb method—that is, on the way out, looking for food, it deposits pheromones on the ground that enable it to retrace its steps. But *Cataglyphis* cannot do this, as the sandy surfaces where they live would just absorb the chemicals. So they have to rely on visual cues, if there are any, and orient themselves in relation to the sun. This they do remarkably well, managing always to make it back to the nest, and by the shortest route, even if they have been on an excursion that has taken them some tens of metres away, at times up to 100 metres.

Desert ants inhabit the Sahara, Mediterranean regions, and the Middle East. Because they live in such arid climes, live prey is in short supply, which is why they go in for necrophagy, feeding for the most part on carrion such as dead insects, other ants, beetles and the like. They supplement this diet by sucking juices from plants.

These 'fascinating creatures', the term used by one of the best specialists on *Cataglyphis*, Rüdiger Wehner, from the Institute of Zoology of the University of Zurich, overwinter in their nests underground. In summer, however, they set out on the hunt for provisions, which they do during the day, unlike other insects, which are active at night. The ants in charge of this provisioning are always the oldest members of a colony; they forage unaccompanied and bring back any prey they find to share with nestmates. This work is not without danger, as their wanderings can expose them to dehydration from the full heat of the sun or to attack by predators. Dawdling is therefore not recommended; they are always in a hurry. This is no doubt why their morphology has developed in a way that marks them off from all other ants. The main feature of this is the length of their legs, which enables them to move much more quickly than normal ants. They also have a peculiar way of holding up their abdomen which, combined with the fact that they are walking on stilts, means they can avoid contact with the overheated ground. When other animals lie quietly in the shade, they go out, like mad dogs and Englishmen, in the midday sun.

Their foraging technique is tried and tested, as has been observed in the field, over time and with great patience, by Wehner and his team. On a recent expedition to southern Tunisia, their aim was to observe colonies of *Cataglyphis bicolor*. On finding a nest, they excavated it so as to select several young workers just out of the cocoon, 'which could be easily recognized by their pale yellowish cuticle'. To identify each ant as an individual, they dabbed colours on its head and thorax, then put it back in the nest. Being nothing if not thorough, they also drew concentric circles on the ground surrounding the nest, so as to be able to make accurate records of the paths followed by each and every ant. Once these parameters were set, the real work of observation could begin; for a period of four weeks, Wehner explains, 'the nest entrance was observed continuously by one person from the beginning to the end of the colony's daily activity period'. Throughout this time, the foragers went about their business visibly unperturbed by the presence of these strange observers.

Five days into the experiment, it was the turn of the first of the painted workers, now much older than before, to join the hunt for provisions. To begin with, they did a quick tour of inspection, venturing no more than a metre or two away from the nest. Then they started to roam farther away in their search for food, each of them setting off in a particular direction. Some came back bearing prey, others without anything.

Before they die, the foragers work on average for only about a week, during which time they go on approximately thirty expeditions. This seems a very short time, yet the observing team are convinced that it is enough for the workers to acquire experience. With each passing day, they venture farther and farther from the nest and spend longer digging for possible prey. If they find nothing during their first forays, they change to a different area for their next ones, whereas those which have had the good fortune to discover a rich source of food keep coming back to it. Gradually, day by day, all the foragers improve their success rate in relation to the number of forays they make.

A memory for images

In navigating their way to their favourite source of food then back to the nest without getting lost, these ants can follow visual cues, if there happen to be any. When their territory contains any scrap of vegetation, for example, they can keep a memory of things seen along the way, such as the position of a bush, patterns of light and shade, or patches of sky criss-crossed or outlined by branches.

There can be no doubt that desert ants have the ability to distinguish between shapes with very similar appearances. This has been shown by Guy Beugnon and his colleagues from the Université Paul-Sabatier in Toulouse. They trained ants belonging to the Mediterranean species *Cataglyphis cursor* to find the shortest way back to the nest through a labyrinth designed out of four boxes linked together. There were two exits from each box, one of which was surmounted by a black design that differed from the design above the other exit only by its shape (circle and cross; star and square; rectangle and triangle; diamond and oval); and only one of these exits gave access to the next box. Amazing as it may seem, it took only a few training sessions for all the navigators to learn, without hesitation or error, the sequence of visual signals pointing to the quickest way back to the nest. Rüdiger Wehner, too, has studied the visual memory of ants, through experiments conducted in the wild, in Greece and Tunisia. He moved a colony of *Cataglyphis* from its natural habitat and transplanted it into an environment of complete and utter desert. Having set out a series of artificial landmarks, he observed the movements of the ants. He explains that they adopted 'a simple strategy' for finding their way home: they tried to link the images of the environment in their memory with those they encountered along their way. This 'landmark map' that the ants use for orienting themselves 'differs considerably from a map used by human navigators'. Wehner concludes that they navigate from one point to another following the cues available along the path used: 'The ant's "map" does not completely cover the area visited during consecutive foraging trips, but consists mainly in the familiar "routes" passing through that area', the routes that the ant will often take when out on a provisioning trip. This, however, is enough for these ants to find their way back to the nest without ever going in the wrong direction.

A compass in the head

Cataglyphis do not rely only on visual cues—if they did, they would be incapable of living in total desert. But even without landmarks, their sense of direction is very good, thanks to a veritable system of automatic piloting in their heads. Their brain, which Wehner compares to a cockpit, is equipped with a battery of navigation instruments, including compasses that tell them which direction to take and a sort of odometer which enables them to measure the distance covered. These instruments are complemented by a 'path integrator' which, rather like an on-board computer, combines the different information supplied from the two other sources and enables the ant to follow the best path.

So, like bees and some spiders, desert ants have a virtual compass in their heads. They have eyes that are able to analyse the polarization of light, a phenomenon produced by the diffusion of sunlight by molecules in the earth's atmosphere which the human eye cannot perceive. Ants, however, can, since their eyes contain receptors which are sensitive to ultraviolet radiation. This has been proved by fitting some navigators with contact lenses that absorb only green light (yes, believe it or not, it has been done), with the result that the ants, deprived of their ability to detect the ultraviolet, lose their way.

As for the variation in the spectrum of light polarization according to the time of day and the position of the sun in the sky, *Cataglyphis* have an answer for that too. Each time they leave the nest, they recalibrate their compasses by making little rotating movements, which Wehner compares to 'graceful little minuets'.

The navigators' odometers, by the way, do much more than measure the distance covered; they also 'calculate' a projection of it on a horizontal plane. This is of practical assistance with all their walking about, since the surfaces that workers walk on are not always flat; they often have to climb up or down bumps in the terrain or negotiate sand dunes. Intrigued by the working of this odometer, Wehner and a colleague, Sandra Wohlgemuth from the University of Berlin, went in for some insect training, laying out a complex and demanding circuit for their ants to run about on. Ants were trained to forage at a feeding site by traversing 'hilly' terrain, simulated by a series of channels sloping alternately upwards and downwards like a switchback. Ants traversing this switchback covered eight metres of terrain, equivalent to a distance in a straight line of 5.2 metres. When they reached the far end of this circuit, the *Cataglyphis* came back to their starting point along a flat surface, which they did in a straight line, covering a distance of about five metres. This proved that they ignored the 'artificial' distance added by the ups and downs. It also demonstrated that they have the ability to factor in the 'third dimension'.

But how can they do this? How can desert ants achieve such feats when all they have is a minute brain that weighs 0.1 of a milligram? This mystery has been addressed by collaborative work among scientists from many different disciplines: entomology, of course, but also neurophysiology, neuroanatomy, behavioural biology, as well as computer science and robotics. They have their work cut out for them; but at least the Wehner–Wohlgemuth experiment described above resulted in the elimination of some of the hypotheses offered as explanation of the amazing ability of desert ants. It had been suggested, for instance, that *Cataglyphis* measured the distance they covered by 'calculating' the energy expended during each outing. Wohlgemuth and Wehner point out that this cannot be: since the workers, by going up and down slopes, covered a distance that was one and a half times longer than what they covered along the flat, logically they should have expended at the very least one and a half times more energy on the way out than on the way back, given that it is much more demanding in calories to climb a hill than to walk along a level surface. So, if the ants were using the energy expended as a way of gauging how far they had travelled, then they would have overestimated the distance to be covered.

Other hypotheses advanced were that they were gauging the distance by their speed or by the time they spent walking. According to Wehner and Wohlgemuth, these hypotheses are no closer to the mark than the first one. The ants' progress was slower when they were climbing hills, and when they were coming down too, than when walking along the flat. This means they took much longer to negotiate the gangway than to return to their starting point. Nor could it be argued that the ants had just followed visual cues, the team of scientists having taken care to rule out that possibility by painting the inside walls of their passageways a uniform grey.

So if all these hypotheses are ruled out, what is left? Wohlgemuth and Wehner take the view that the secret lies in the actual motions made by an ant as it walks. The swaying of its body, they argue, is what supplies the information necessary to the proper working of their odometer. One thing that is known for certain about ants is that they are affected by the force of gravity. They can probably also gauge the angle of inclines with proprioceptors, receptors sensitive to their movements, located between the different parts of their bodies, between the head and the thorax, the petiole and the abdomen, or in the joints of their legs.

None of this, however, says anything about how an ant's brain integrates these different parameters. It explains neither how the navigators 'calculate' the distance covered during their wanderings, nor how they find their way back to the nest without ever straying off course. The desert ants' automatic pilot remains a mystery.

16
Honeypots

In general, insects' lives are governed by the supply of available food. In temperate climates, the adults develop in the spring and summer, a time when there is no shortage of prey or edible vegetable matter. Then the females lay and soon afterwards die. The next year's generation spends the winter in a latent state, in the form of eggs or larvae which will not mature into adults until the following warm season. And so the whole cycle begins again.

When it comes to ants, however, which live for several years, this system does not work. The improvident cicada in La Fontaine's fable, singing all summer then having to beg food from her neighbour the ant, was knocking on the right door, as ants spend the warm season laying in supplies not against the coming winter but against the following springtime, when the colony reawakens.

This business of stocking foodstuffs is beset with potential problems. Supplies are precious and must be protected against possible spoilage and the danger of theft by other insects. Seed-eaters, for the most part, have come up with a simple solution: well inside the nest, they set aside a chamber reserved for storing seeds. Many carnivorous species, on the other hand, do things very differently, as was first described by Peter Nonacs of the University of California (Los Angeles): they just use their offspring as a reserve food supply. When there is nothing left to eat, the queen and the workers start feeding on some of the colony's larvae.

The smartest solution to the problem must be that invented by honeypot ants: they use some of their own workers as a larder. This is a good way of adapting to the environment, because ants of the genus *Myrmecocystus* live in arid and semi-arid areas of Australia and America, where food and water are plentiful only during the brief rainy season. At other times of the year, these necrophagic ants can still find some dead insects. But what they lack is sugar, as they can no longer collect nectar from flowers or honeydew—what a nice word for what is in fact the excrement of aphids! Accordingly, some of the workers take advantage of the times of plenty to gorge themselves on sweet things, stuffing themselves until their abdomens swell like balloons. In general, it is the largest workers which turn themselves into these 'honeypots' (also called 'repletes'). They start specializing in this practice right at the beginning of their adult lives, when their bodies are still soft and flexible, though in young colonies, according to the head keeper of the insectarium at Cincinnati Zoo, Randy C. Morgan, 'it is not uncommon to find small workers serving as somewhat ineffective repletes, with their tiny abdomens distended to bursting point'.

When the dry season starts, these walking barrels settle in the deepest part of the nest. They hang in groups from the ceiling of a chamber, 'literally imprisoned by their globose abdomens ballooned to the size of small grapes', as Morgan puts it. He gives a striking account of an expedition he led to a desert site near the Chiricahua Mountains in south-east Arizona, the aim of which was to acquire a complete nest for the Zoo. Working at night and after having dug for hours to get at the deepest parts of the underground nest, he and his team eventually came upon the honeypots: 'Deep within their subterranean nest, honey ant repletes hung from chamber ceilings in golden clusters. In the interplay of our flashlights the repletes sparkled and glistened like living jewels.' The many onlookers who had turned up to see what all the digging was for were much impressed by the sight, which made up for the oppressive heat and choking dust.

If no myrmecologist comes along to dig them up, the honeypots hang there in their obese huddles. Then, one after another, when the members of the colony are close to starvation point, they start to regurgitate their reserves of sugar and water.

Confectionery

Myrmecocystus have perfected their way of storing a reserve of food. One might suppose that they spend carefree winters, well inside the nest, gradually drawing on the supplies laid down over the previous summer. But life is not quite so easy in these nests. Nest-raiding by colonies of the same species is not infrequent. The invaders make off not just with any repletes they find in the nest but also with eggs and larvae which they will take back to their own nest as slaves. It must be added that the American badger and other denizens of the desert find the swollen honey-filled ants very much to their taste and are known to dig deep to get at the sweetness they crave.

Nor are animals the only nest-raiders. Until the early years of the twentieth century, Mexicans and native Americans also looked on honeypot ants as sweet supplements to their diet. They would excavate nests and take out the repletes: 'Sweet-toothed human predators typically hold a replete's head and thorax with the fingers, bite off or rupture the fragile abdomen, then suck its contents into the mouth' (Morgan). This practice seems to have died out since confectionery has become readily available in shops. Man-made sweets 'not only taste better but are much less work to procure,' says Morgan. This certainly makes for a safer life for honeypot ants.

Part IV
Advantageous Liaisons

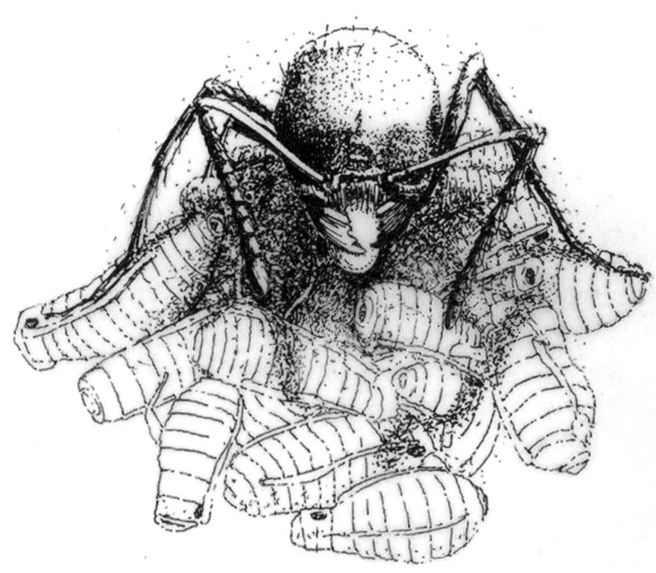

Drawing 4 Raising aphids Some ants live in symbiosis with aphids, feeding on their secretions and caring for them.

17
Colonies and their livestock

Wherever ants live, their very presence is welcome to many animals, first and foremost among these being of course their predators, which exploit ant colonies as a plentiful source of food. Then there are the arthropods and vertebrates which join the raiding columns of army ants like camp-followers, finding this situation to their advantage whether for moving to a different place or for finding nourishment. They, however, are mere opportunists.

Other animals too, isopod crustaceans, for instance, but especially many insects, contrive to share the comfort and sustenance afforded by an ants' nest. These free-loaders must make sure they are accepted by the colony; and to this end, they have adopted various stratagems. One example is morphological mimicry, a ploy used by a good many spiders, bugs, and beetles which have quite simply reshaped their anatomy, 'disguising' themselves as ants. Or there is the tiny wasp *Lepidopria pedestris*, which as soon as it manages to enter a nest of the housebreaking ants *Solenopsis fugax* loses its wings. In this way, it looks more like one of the ants, being no larger than they are; and according to Luc Passera and Serge Aron in their book *Les fourmis: comportement, organisation sociale et évolution,* it is 'only by its colour, a little darker than the ants, that a skilled observer can recognize it'.

To become members of a colony, some species mimic behaviour rather than shape. A case in point is the pupa of the *Hyalymenus* bug, which imitates the zigzag gait of workers. To be accepted by ants, chemical signalling is another good method to use. Some spiders adapt their sense of smell the better to detect the pheromones produced by workers, thus allowing them to migrate at the same time as the colony. Some coleoptera, like staphylinid beetles, are even more ingenious: in early winter, when an adult ventures into a nest of *Formica polyctena* wood ants, it raises its abdomen to uncover its appeasement glands which secrete substances that have the immediate effect of reducing the aggressiveness of the workers. Simultaneously, it will touch any hostile ant with its antennae, while producing other so-called 'adoption' molecules which will result in its being carried into the nest by the workers.

However, when it comes to duping workers, first prize must go to 'chemical camouflage', a technique that fly larvae of the *Syrphidae* family have got down to a fine art. They manage to get themselves carried into the nest by workers that live there; and once they are in among the brood, they happily set about eating it. They have of course adopted a good stratagem to outwit the vigilant workers, which is to pass themselves off as ant pupae by producing the very same odorous hydrocarbons as they do. This fools the workers into treating the young flies as though they belonged to their own brood. Or take the little beetle, *Martinezia duterteri,* which lives in the nests of various ant species and frequently moves in with other colonies in different nests, rubbing against the workers to take on their smell. Evidence of this is seen when it enters a new nest: at first the beetle is attacked, but after a few days, it gets adopted.

Whichever trick they try, these spiders, bugs, and beetles all end up by being accepted into a colony, taking up their quarters either in the nest or close by. Once they are inside, the more discreet among them are content to live on any leftovers from the workers' diet. Some are smart enough to live off food regurgitated by the ants. But the carnivores eat their fill of the colony's brood, as its eggs, larvae, and pupae are too fragile to defend themselves; and some just set about eating the workers' abdomens.

So some of these free-loaders have in fact become parasites, behaving like cuckoos in the nest, or predators that could not survive in the world outside the nest. The ants accept them only because they do not recognize them for what they are. Nor do they benefit in any way from their presence in their midst.

Aphids as milch-cows

There can, however, be privileged relationships between ants and other insects. In such cases, there is a genuine association of benefit to both partners; each of them thrives and can even become dependent on the presence of the other. When that happens, ants and insects live together in full symbiosis.

This cohabitation of species, known as mutualism, is a common practice in the Earth's ecosystems. No exception to this rule is made for human beings: our digestive tube harbours a multitude of bacteria which are indispensable to our metabolism. Similarly, there are cases of mutualism, which are in their own way highly successful, between ants and a variety of insects such as aphids, mealy bugs, leaf-hoppers, and other homoptera. In such associations, called 'trophobiosis' (from the Greek stem *tropho-*, meaning 'nourishment, to nourish' and *bios*, 'life'), the ants draw sustenance from the secretions of the insects (the 'trophobionts'), who benefit in return by being well cared for. In a way, the ants are practising a form of animal husbandry, establishing with their trophobionts a relationship not very different from that between human farmers and their livestock. This highly peculiar behaviour was noticed by the Swiss naturalist Pierre Huber: in 1810, in his *Recherches sur les mœurs des fourmis indigènes* (Studies of the customs of indigenous ants), he compared the ants to 'shepherds' looking after 'their cattle and goats'. Some decades later, another Swiss, Auguste Forel, took the comparison farther, calling aphids the 'milch-cows' of ants.

This is an apt metaphor, though the 'milk' produced by the ants' livestock is in fact honeydew. Bugs and homopteran insects feed on the sap of plants which they digest before excreting the surplus from their anus in the form of sweet drops. However unappetizing this may appear to us, ants are not the only ones who enjoy this peculiar type of 'honey'. The manna from heaven which the Old Testament says was given as food to the children of Israel was very likely made of the droppings of mealy bugs; and to this day, Australian aborigines gather and consume honeydew or what is called 'sugar-lerp'. Besides, the honey we all enjoy is in large measure made from honeydew picked up by bees from the surfaces of bushes and trees and is therefore nothing other than insect droppings processed through the digestive tubes of other insects. Anyway, this honeydew is a boon to ants, since they draw from it all the nutriments they need: sugars in large amounts, as well as amino-acids, B vitamins, and minerals.

Ants that live in symbiosis with homopterans do not just collect the honeydew; they actually 'milk' their aphids by tapping them gently with their antennae and legs; and in response to these stimuli the trophobionts secrete their tasty drops. Workers of *Formica lugubris* spend almost half their time tending to a group of aphids which do not excrete their honeydew until they are stimulated to do so in this way. In other cases, the trophobionts 'summon' the ant by showing an anal droplet then reabsorbing it several times, which alerts the ants. If the ants are not interested, the drop is not evacuated but is retained inside the digestive tube.

For ants, the advantage of this association is obvious: thanks to their livestock, they have on tap an important source of sustenance. Surprisingly, too, the 'domesticated' aphids are much more productive than their untamed fellows. For example, some species of Mexican leaf-hoppers of the genus *Dalbulus* cohabit with ants and secrete three to six times as much honeydew as other undomesticated leaf-hoppers. The fact is that, for aphids, leaf-hoppers, and other homoptera, there is nothing compulsory in this association with ants. Some of them get along quite nicely without any such association. Nevertheless, those which, through evolution, have turned into trophobionts benefit greatly from it.

Hygiene, comfort, and protection

For one thing, the aphids' hygiene is much improved, since their ant associates help them to keep clean. Aphids have to get rid of their droppings lest the sugars ferment and turn into mould, which is not good either for the insect or for the plant in which it lives. Homoptera leading an independent life can of course give themselves a good shake, thus throwing off the drops of excrement. But those that have become trophobionts do not have this problem and do nothing to rid themselves of their honeydew. They actually present their liquid in a way that enables the ant to absorb it more easily.

The greatest benefit that homoptera draw from the presence of ants is that it affords them comfort and protection. Workers are highly skilled in ways of keeping their dependants away from unfavourable weather conditions and out of danger. They also look after their subsistence by erecting shelters round the plants they feed on. European red ants, for instance, build protective earthworks round the herbaceous stems colonized by aphids. Some ants from the rain forests of Malaysia make underground chambers for their mealy bugs and regularly move the insects to keep them in contact with roots. As for *Lasius neoniger*, they adopt the eggs of the American corn root aphids with which they have an association. In winter, they keep them inside the nest and house them with their own brood. When springtime comes, they carry the young larvae from the nest to a nearby root, taking care to put them on the plant they prefer.

Like any conscientious breeders of livestock, ants defend theirs against their enemies. Any predators of aphids, and the minute wasps or parasitical flies that lay their eggs in the bodies of insects, are very smartly seen off the premises. The effectiveness of this sort of bodyguard is attested by studies done on *Syrphidae* in the orange groves of Algeria: the voracious larvae of these flies can destroy almost 60 per cent of aphid colonies unprotected by ants, but only 12 per cent where there are ants on patrol.

However, not all trophobionts are so well served, as some of their parasites or predators have hit on a counter measure: they camouflage themselves to evade the ants. To keep on eating the aphids that are its staple diet, the larva of the green lacewing contrives to outwit *Camponotus* ants. Like a wolf in sheep's clothing, it hides under a tuft of waxy filaments derived from its victims, and so it is extremely difficult, even for the trained eye of the entomologist, to make out the lacewing among the aphids. Even the ants are fooled by this ploy.

All things considered, herbivorous insects that live in close proximity to ants generally benefit from the association. Thomas Flatt and Wolfgang Weisser, formerly at the University of Basel, established that aphids which live close to *Lasius niger* ants are not only more prolific but have a longer life and greater reproductive success.

1 Two major organizations of ant societies. **a**. In *Monomorium sydneyense*, all workers are daughters of a single queen, which lays all the eggs in the colony. **b**. In *Pheidole desertorum*, several queens share reproduction, which creates different lineages of workers within the same colony.

2 Worker size polymorphism. *Pheidologeton* marauder ants have one of the largest size disparities between the largest and smallest worker ants in a colony. The ants pictured here, large and small alike, are all sisters. Supermajor workers have powerful mandibles that are useful for cutting up and carrying large prey items such as this beetle grub.

3 Different brood states **a**. Minor workers of *Pheidole desertorum* tending eggs and larvae. **b**. Large cocoons of *Lasius nearcticus* house pupae destined to become queens, while small cocoons produce workers and males. **c**. Brood inside the brood chamber of a large *Iridomyrmex* colony. **d**. Ants undergo complete metamorphosis much like butterflies. Here in *Aphaenogaster picea*, the grub-like individual on the right is a final-instar larva, the other two being pupae. As pupae approach maturation they turn darker in colour.

4 The different phases leading to the formation of a new colony **a.** A winged *Formica* queen leaves the nest for the mating flight. **b.** Large mating flight of the leaf-cutting ant *Acromyrmex versicolor* in Arizona. **c.** A *Formica paralugubris* queen mates with a male. **d.** A *Pheidole* queen cares for her brood in her claustral nest-founding chamber.

5 Ants attack a great variety of prey. **a.** *Leptomyrmex* workers co-operate to bring a grasshopper carcass back to their nest. **b.** *Amblyopone australis* ants are subterranean predators. Here they are seen attacking a moth larva. **c.** This *Odontomachus* trap-jaw ant has caught a termite. The odd shape of the ant's head is due in part to the extensive musculature behind the trap-jaw mechanism. **d.** A *Myrmecia* worker carries her prey, a wasp, back to the nest.

6 Different types of ant nests. **a**. A typical thatch mound nest of *Formica* ants. **b**. Arboreal *Tetraponera* ants nest in hollow spaces in twigs and branches. The ants' slender bodies allow them to manoeuver in narrow spaces. **c**. Workers of the army ant *Eciton burchellii* create a bivouac in the middle of which the queen is protected. **d**. *Azteca* ants living in cavities of a *Cercopia* tree.

7 Army ants. **a**. Soldiers of *Dorylus molestus* guard an emigration column. **b**. *Eciton burchellii* workers forming a living bridge across a gap in their path. **c**. A recently hatched male of *E. burchellii* walks in an emigration column. Notice a myrmecophilous beetle riding on his thorax. **d**. Details of the head and mandibles of an *E. burchellii* soldier.

8 Weaver ants. **a.** Winter leaf-drop reveals a number of old *Oecophylla* tree-ant nests. These nests all belonged to a single colony. **b.** *Oecophylla smaragdina*, the green tree ant of Northern Australia, make their nests by tying living tree leaves together using silk produced by their larvae. **c.** Major and minor green tree ants standing on top of their silken arboreal nest.

9 Honeypot ants like *Myrmecocystus mexicanus* have an unusual food storage system. Some repletes become engorged with food and hang from the ceilings of chambers deep underground until the dry season when they can feed other members of the colony.

10 *Myrmica* ant tending aphids.

11 Leafcutter ants. **a**. *Atta cephalotes* workers characteristically slice circular patterns in leaves. **b**. Workers bring home the harvest. **c**. The fungus garden within the nest. **d**. Leafcutter ants are sometimes referred to as parasol ants because of their way carrying their cargo.

12 Throphallaxis between two workers in *Oecophylla smaragdina*.

13 Some ant species are organized into supercolonies, allowing brood and worker exchange between different nests. **a**. In the polygynous form of *Solenopsis invicta* nests are frequently interconnected. **b**. The wood ant *Formica paralugubris* can form populations containing large networks of interconnected mounds as indicated by this group of nests each marked by a yellow balloon.

14 Invasive ants. **a**. Workers of the ghost ant *Tapinoma melanocephalum* drink from a drop of water **b**. A male and workers of the red imported fire ant *Solenopsis invicta*. **c**. A queen, a male (with wings) and workers of the pharaoh ant *Monomorium pharaonis*. **d**. Two workers of the Argentine ant *Linepithema humile* exchange information.

15 Unusual family structures. **a**. New queens in *Cataglyphis cursor* are produced by parthenogenesis. On this leaf a worker (right) is near a queen. **b**. In some *Pogonomyrmex* species, queens mate with males of their own lineage and males of another lineage to produce new queens and workers, respectively. **c**. In the little fire ant *Wasmannia auropunctata*, queens produce new queens and males by clonal reproduction. The workers, pictured here, care for the brood and the queens

16 Robots developed to study collective behaviour **a**. Swarmbots form a chain to cross a gap in their path **b**. Khepera robots mimicking ant foraging behaviour **c**. Communication using light in a group of E-puck robots. **d**. Tiny (2-cm) Alice robots forage for large and small objects. Large object foraging requires co-operation between the robots.

Nomads and shepherds

In the early 1980s, Ulrich Maschwitz and Heinz Hänel of the University of Frankfurt discovered one of the oddest couples: a Malaysian ant of the genus *Dolichoderus* and the plant-sucking louse *Malaicoccus*. The eccentricity of these ants is such that they have become veritable shepherds, living a nomadic life, moving on whenever they need to satisfy the feeding needs of the flock that gives them their livelihood. The workers cart their bugs about either in their mandibles or on their backs, for ever transporting them to greener pastures—in other words to plants in their full growth, full of sap rich in protein.

Just as nomadic as army ants, *Dolichoderus* never make a permanent nest, preferring to make temporary bivouacs out of their own bodies, which they form into a compact mass to protect their trophobionts. During these halts, ants and lice live on equal terms, the bugs and their larvae being usually mixed in among the ants' brood. Once food starts to run short in the vicinity of the bivouac, or if conditions of temperature and humidity require it, the whole shebang has to up and move house. So camp is struck and the colony—that is, the queen, about 10,000 workers, 4,000 larvae and pupae, plus a good 5,000 bugs—takes to the road. The workers carry their own brood as well as their trophobionts, pausing along the way to set down their charges and let them sample the surrounding vegetation. If the *Malaicoccus* start to perforate the leaves, showing that they find them to their taste, this means the site is suitable. In this case the shepherds bring up the rest of the flock and set up new quarters. There is no regularity in this process of moving house; it happens only when the flock require it. Over fifteen weeks of observation of colonies, the Frankfurt team noted that each month the colonies might move either once or twice, or not at all.

Milk and meat

Whereas *Dolichoderus* ants care for their trophobionts so as to consume their 'milk', other types of ants raise flocks for their 'meat'. This is the case with *Formica paralugubris* wood ants, which consume 30 per cent of their aphids, a fact established by the Swiss entomologist Daniel Cherix. Although African *Melissotarsus* ants live in association with mealy bugs, the latter produce no honeydew, but are kept in the nest and painstakingly looked after, so as to be eaten and thereby supply the proteins required by the ants.

The diet of *Lasius niger* is mixed, with proteins and sugars, and the ants adapt their behaviour to this nutritional requirement. When they lack sugars, they harvest honeydew from aphids; when they are deficient in proteins, they eat the actual homoptera. The size of the flock also has a bearing on this: if the ants have enough aphids to keep themselves well supplied with sugars, then they kill more of them to eat. *Lasius niger* actually have associations with two aphid species which they use indiscriminately as a supply of honeydew or as prey, depending on which one they happen to find first. From this it can be seen that there is no hard and fast boundary between mutualism and exploitation.

The emergence of trophobiosis is not in any sense a recent development, as is shown by pieces of amber from the Baltic dating from the early Paleozoic Era that already show aphids associated with ants. To begin with, the herbivorous insects were probably only prey for the ants, which must have been attracted by the sugary excretions and begun to consume honeydew they found on leaves. Mutualism, whether obligatory or not, then gradually developed, the ants selecting as associates the slowest-moving aphids which most resembled their own larvae, and eating the others. Eventually the two partners evolved together—'co-evolution' is the term used—, each of them adapting its morphology and behaviour to the needs of the other. Thus aphids, once they are protected by ants, simplify their defence tactics. When unprotected by ants and attacked by predators, they emit an alarm pheromone that stimulates their fellows to flee or drop to the ground. However, such alarms do not affect trophobionts: they do not move, but rely on the ants to see off the enemy. As for scale insects of the *Diaspididae* family, which serve as prey for their associated ants, when touched by the latter they stop secreting the silky armour that usually protects them against enemies.

Honey glands in caterpillars

The caterpillars of many *Lycenidae* and *Riodinidae* butterflies have also established solid links with ants; and the morphological changes this has induced are even more astonishing. On the face of it, these caterpillars would appear to have nothing to offer ants, for they consume neither sap nor vegetable matter. So as to attract workers, they have adapted by acquiring a 'honey gland' which secretes a sweetish liquid rich in nutrition. To avoid being confused with mere prey, they have also covered their bodies with rather special glands which produce substances that probably have the role of calming the ants and reducing their aggression.

For the caterpillars, such adaptations require great expenditure of effort in terms of energy. In exchange, though, just like the aphids and mealy bugs, they benefit from the protection of the ants, which defend them against their parasites and enemies. So they too derive great advantage from this association in terms of development, survival, and reproduction. In Australia, there is even a male *Lycenidae* butterfly that obtains a mate by identifying the ants accompanying the females; and it is also the ants that enable the females to find the best plants on which to lay their eggs—one good turn always did deserve another.

In this association, too, mutualism can develop into parasitism, though in this case it is the trophobionts that take advantage of the ants. The caterpillars of *Maculinea* butterflies have hit on a good ploy for self-advancement. During the earliest stages of its development, the butterfly larva lives and feeds on shrubs of thyme and oregano. At the end of summer, it drops to the ground and waits concealed among tufts of grass until a *Myrmica sabuleti* worker turns up. As soon as it feels the ant touching it with its antennae, it attracts the ant by secreting a sweet liquid from its honey gland and twisting its body in a weird way. The ant is completely taken in by this, adopts the caterpillar, and carries it back to its nest. Once inside, the intruder spends the winter there and, as soon as spring comes, changes its eating habits by turning carnivorous and devouring its hosts' brood. There are only a few European species of *Maculinea*, such as the Alcon Large Blue and the Mountain Alcon Blue, which do things differently: their caterpillars behave more like cuckoos, in that they only require feeding by the workers.

In tropical Asia, some *Maculinea* use a different tactic to exploit ants. The adult butterflies settle among aphids or mealy bugs, consume the honeydew, then lay their eggs. Once they have hatched, the caterpillars develop not only by enjoying the sweet liquid but by eating some of the homoptera. In this case, the caterpillars win on all scores, being apparently the only ones to benefit from the association. Why the ants tolerate the *Maculinea* is not known.

Between perfect mutualism and pure parasitism there are many intermediate degrees of relationship. A mode of association that grows up between two species may also vary during evolution. Ants have formed many alliances with other insects, some of which are of immense benefit to them and in others they are taken advantage of. Such is the risk in any partnership.

18
Ant trees

It may seem quite plausible that ants should have made indispensable allies of aphids, mealy bugs, and butterfly caterpillars. That they have also entered into associations with plants, to the point of actually living in symbiosis with them, on the face of it appears much more bizarre. However, it is a simple fact that mutualism between plants and insects does exist; indeed, it is very frequent in tropical habitats, where there are trees known as 'ant trees'. Co-evolution has resulted in symbiotic alliances between some plants and insects which are among the weirdest and the most sophisticated known in the natural world.

The benefits to ants of any such association are obvious, since it affords them what Luc Passera and Serge Aron call 'bed and board'. Some tropical trees, such as the Cecropia, or trumpet tree, of the Central American rain forests, or the Macaranga of South-east Asia, make excellent living quarters for ants, because their stems and internodes are full of cavities which colonies can easily take over. The workers build their nests in them and do not hesitate to enlarge any cavities that are too small for their purposes, even going so far as to tunnel through the ends of the branches to make multiple exits.

These dwellings may not always offer all mod cons, but the ants try to cope with any inconveniences. For instance, *Cataulacus muticus*, which live in the hollow internodes of giant Malaysian bamboo, have to face regular flooding brought on by tropical downpours. They try to prevent the water from entering the nest by stopping up the entrances with their heads. If that does not work, then, as Passera and Aron put it, 'the workers drink up the excess water, go out onto the stem of the plant, hold their abdomen vertical, and expel the droplet that forms at the end of it'.

In Africa and Central America, the thorns of acacias, especially the variety known as bull's-horn thorn, or swollen-thorn acacia, make a good home for ants. The thorns grow in pairs along the branches, which explains their name, and have a hard surface covering a pulpy interior. This makes an ideal home environment for ants. In Guatemala, where there are acacias with two different sorts of thorns, workers of *Pseudomyrmex ferrugineus* are careful to install their queen inside the stronger variety, thus protecting her from birds.

Not that there is anything unusual per se in ant colonies inhabiting convenient trees. It happens the world over, in European forests for instance; many ants are known to build nests in tree trunks or stems, the actual species of plant being a matter of indifference to them. If it suits their purposes, they will colonize it. When, however, trees become associated with a particular species of ant, they do not just give them houseroom, they also feed them; and this they do by growing organs which appear to have no use other than to contribute to the well-being of the ants. Acacias, for example, as well as many other flowering plants, have glands that produce a sweet liquid. These 'extra-floral' nectars serve not to attract pollinating insects but only as a treat for ants. The same can be said of the nutritive packets of corpuscles located at leaflet tips—called 'Beltian bodies'—which are rich in proteins and lipids and which the ants can easily gather. Species of the genus *Pseudomyrmex*, the main partners of acacias, can thus find all the nourishment they need on their doorstep. Macaranga and Cecropia also have nutritive bodies exclusively reserved for their ants; and if they are deprived of their favourite insects, they either produce fewer of these nutritive bodies or else cease production altogether.

Security guards

The plant may expend energy to produce its various nourishing organs, but this is to its ultimate benefit. Ants may not be good lenders, as La Fontaine's fable suggests, but this does not mean they are complete egoists, especially as some give-and-take is in their own interest: in exchange for their bed and board, they protect the plant against its enemies, in particular herbivorous insects which nibble away its leaves.

Central American acacias find that their *Pseudomyrmex* act as formidable security guards. These ants, which are quite large and have a fine sting, patrol by the thousand among the leaves and are ruthless with intruders. Any interloper is instantly attacked by workers, who take hold of it in their mandibles and sting it, usually putting it to flight. If it persists, they release an alarm pheromone to recruit nestmates which then arrive in great numbers to see off the intruder. There can be no doubt that the patrols are effective: fewer than 3 per cent of acacias associated with ants harbour insect pests, whereas almost 40 per cent of those which have not been colonized by *Pseudomyrmex* do have pests.

Even small ants manage to protect the plant that houses them. Tiny as they are, *Petalomyrmex phylax* contrive to defend the rainforest tree with which they are associated (*Leonardoxa africana*). Its mature leaves are too tough to be attacked and need no particular protection. The immature leaves, however, are at risk of being eaten by the herbivorous insects which lay their eggs on them. And that is where the workers come in. It is also where the plant keeps its ants by secreting a substance that attracts them. At blossoming time, on the other hand, the plant releases chemicals that have the effect of making the ants leave, which they do for just long enough to allow bees in to pollinate their host.

Leonardoxa africana is not the only plant to communicate with insects through chemical messages. If Macaranga and some Cecropia are attacked, they too produce chemicals as alarm signals. This serves to alert the ants, which then scurry in droves to the aid of the damaged leaf and rid it of the pest attacking it.

Ants living in symbiosis with trees are combative—in fact, they are among the most aggressive of all ants—and do not hesitate to attack enemies much larger than themselves. In West Africa, *Tetraponera aethiops* ants, which live in the hollow stems of a small tree of the Passifloraceae family, do not just rid the plant of the caterpillars of butterflies, moths, and beetles. They also keep Colobus monkeys away, by stinging them. *Crematogaster* ants even manage to protect their acacias against giraffes, which are partial to the plant. At the slightest vibration of the thorns in which they live, the ants mobilize against the aggressor, biting it and putting venom into the wounds they inflict. Young giraffes have no liking for this treatment and do not stay long in the vicinity.

Devil's gardens

Ants take their protective role very seriously and also liberate their tree or shrub from any other plants that might steal its sunlight or hinder its development. This was first observed by an

American ecologist, Daniel Janzen, in the 1960s. In Costa Rica he found that the only Cecropia which were not encumbered with climbing lianas were those housing colonies of *Azteca* ants. When Janzen twined some tresses of a liana about one of the trunks, they were instantly bitten by workers and so did not survive for more than a few days. He tried different experiments in Mexico, depriving certain acacias of their *Pseudomyrmex*. The immediate result was that the trees started to wither away, whereas those still protected by the ants continued to thrive. Patrolling ants not only attack any Coleoptera such as Colorado beetles, or butterfly caterpillars that trespass on the leaves, they also chew and pull to pieces any foreign plant that happens to grow at the foot of the tree.

In the Amazonian forests the pruning and weeding done by *Myrmelachista* ants round their trees is even more striking. They regularly climb down from the nest, not to feed but for the sole purpose of destroying any plant that dares to grow within a radius of a few metres from the trunk. They are a dab hand at it too: they attack a young leaf at its most sensitive point, at the base of the vein supplying its nourishment. They begin by biting the leaf; then they turn round and puff a jet of poison from the end of their abdomen into the wound. This substance is a genuine herbicide and will result in necrosis that rapidly spreads along the leaf before affecting the whole unwanted plant, which can thus be eliminated very smartly. Further proof of this was supplied recently by Megan E. Frederickson and Deborah M. Gordon of Stanford University: having planted cedars round the favourite tree of a colony of *Myrmelachista schumanni*, they saw the ants immediately flock to the intruders and inject their venom into the leaves which, less than twenty-four hours later, had shrivelled. The only cedars to develop normally were some which had been treated with insecticide.

With their good access to sunlight, ant trees will thrive. They soon send up shoots which the ants take over as branches of the main nest, surrounding them with a buffer zone one to three metres in width. This explains why some areas of the Amazonian forest are strangely inhabited by a single species of tree, in clear contrast with the prolific varieties growing in areas close by. People living in these areas called them 'devil's gardens', following the traditional belief that they were the haunt of an evil forest spirit. We now know that the devil lies in the detail of the herbicide used by ants to protect their favourite tree.

Very different sorts of gardens are tilled by *Camponotus* ants to produce the flowers with which they have an association. In the branches of any sort of tree or shrub they assemble clumps of earth, detritus, and vegetable fibre. They then carry up seeds of their favourite plant which grow well in this humus, putting down roots that become part of the structure of the garden. In return, the plant provides the *Camponotus* with its nutritive bodies, the nectar from its flowers, and the pulp of its fruit.

Germination and pollination

Even when not engaged in tending their own gardens, ants play a significant role in disseminating seeds: it has been estimated that more than 3,000 plant species, especially some common in Africa and Australia, derive benefit from this. Their seeds could appear to have been designed for the insects that spread them, with excrescences rich in lipids which attract the ants in the first place while providing handles that make them easy to carry about. Once they have got the seeds into the nest, the insects eat off the excrescences but discard the seeds which, being still intact, can then germinate. It is very much in the interest of plants that workers play this role in spreading their seeds. As gatherers, ants are very able and nimble, much quicker than other granivorous animals, which enables them to protect seeds against being pecked by birds or gnawed by rodents. In South African and

Australian bushlands, ants bury seeds close to the nest, which saves them from being burnt in bushfires. And of course ants' nests and their environs are a milieu that is good for germination.

Workers can also serve as pollinators for flowering plants. Most species of ants are in fact not suited to this task, since their metapleural gland secretes toxic substances. These are certainly useful for killing pathogenic bacteria infecting a colony; but just as they destroy the membranes of micro-organisms, so they also damage grains of pollen. *Camponotus* ants living high in the Sierra Nevada in California, a habitat too arid for pollinating insects, have adapted to this situation. Having no metapleural gland, they have become very good at dispersing pollen.

In some cases, the plants themselves adopt a particular morphology so as to exploit ants for their own advantage. This can be seen in some orchids, whose shape appears to have been expressly designed so as to make their pollen stick to the ants' foreheads, well away from the harmful secretions. Other flowering plants have devised different tricks: they attract male ants who have no metapleural gland by releasing imitations of the sex pheromones produced by female ants. This is a good way of duping the insects into believing that they are mating with a series of young queens when they are actually carrying pollen from flower to flower.

Close partnerships

It is not clear how such special relationships between insects and plants could have arisen, especially when they result in an inseparable couple made of a species of ant and 'its' plant. One thing, however, is certain: these partnerships, just like those between ants and trophobionts, are the products of co-evolution.

We may hypothesize that, in the beginning, the ants were merely drawn to the nectars produced by some shrubs to attract pollinators. With the aim of monopolizing the resources, they took to chasing away any other insects that trespassed onto the plant. This being of benefit to the plant, it developed extrafloral nectars and nutritive bodies to retain the services of its protectors; and they, now that they had a plentiful supply of provisions, took up residence in the plant.

In tropical forests, it may have been competition between different species of tree-dwelling ants that made some of them develop an association with a particular tree. Species that came off second best and were evicted from the canopy would have had to build their nests in the crowns of other trees which, finding the presence of ants advantageous, adapted to the needs of the insects.

However they came about, once these close partnerships were formed, it was in the interest of both parties to remain united, since they both found them extremely beneficial.

19
Attines and fungus getting on famously

In Guadaloupe and throughout much of Latin America they are called parasol ants; and when you see them making for their nest holding a large leaf over their heads, you could be forgiven for thinking they are trying to keep the sun off. However, these leaf-cutting ants carry their sunshades even when foraging at night, their sole purpose being to harvest vegetable matter as fertilizer for their gardens. Leaf-cutting ants of the genus *Atta*, found in Central and South America, mostly feed on fungi, which they grow with great care on the leaves they have collected. This has led entomologists to claim that agriculture was first invented by ants. It is true that fungus-growing (attine) ants started to cultivate their fungi about fifty million years ago, well in advance of *Homo sapiens*, who discovered the possibility of sowing and reaping only about 10,000 years ago.

Here we have another example of real symbiosis of immense benefit to both parties. The interest of the fungus is served by its being protected against fungivores and parasites. And for *Atta* ants the advantages of this mutualism are enormous, as they have a staple diet of fungi, mainly basidiomycetes, which their larvae consume almost exclusively.

The dependence of the ants on the fungus did not just spring up overnight. The original fungus-growing ants were not leaf-cutters, but debris collectors, using bits of withered plants for cultivation of a relatively unspecialized mycelial fungus that could live independently from the ants. Nowadays, most of the more than 200 species of fungus-growing ants still use leaf-litter debris for fungal cultivation. These species are much less conspicuous than leaf-cutting ants. They typically form small colonies comprising a few hundreds workers and they lay out their little gardens inside pretty unsophisticated nests, which are dug very close to the surface or even protected by just a few stones. Though not much larger than a golfball, such a garden supplies all their needs.

The evolutionary transition from debris-collector to leaf-cutter was accompanied by a dramatic increase in worker numbers and complexity of social organization. Some extant leaf-cutting nests are estimated to live for ten to twenty years, contain five to ten million workers and maintain 500 football-sized fungus gardens. In these extended underground nests, as big as a bus, work is highly organized and performed on the model of an assembly line. Large workers go out to gather vegetation of all kinds, leaves, fruit, blades of grass, whatever suits their species, and bring it back to the nest. There they dump their loads of vegetable matter, which is cut up into tiny fragments by smaller ants. Then comes the turn of even smaller workers who crush and mould the plant debris into damp balls. The end result of this sequence of operations is that vegetation is turned into a spongy paste consisting mostly of microscopic leaf fragments and plant juices. This is how, like Voltaire's Candide, workers cultivate their garden in which the symbiotic fungus will grow. When it starts to appear, they tend it with great care. They weed out any hyphae that are growing too slowly, then replant these filaments of mycelium in newly prepared gardens containing vegetable paste. They also regularly lick the surface of their crop to rid it of any moulds of foreign vegetable species.

Leaf-cutting ants are therefore, for the most part, genuine farmers. They set up their garden plots right in the centre of the nest, where the queen is quartered and lays her eggs. This is convenient for the larvae, as the nourishment they need for growth is right beside them. And when the time comes for the young queens to fly away, they will carry in a pocket inside their mouth cavities some filamentous strands of hyphae which will enable them to start a new garden for their own society-to-be.

Parallel evolution

Leaf-cutting ants of the genera *Acromyrmex* and *Atta* do not just cultivate fungi; they have really domesticated them. By replanting strands of hyphae on their carpet of vegetation, they prevent sexual reproduction and spore formation in their basidiomycete cultivar. Spores are superfluous to their needs, as they do not eat them, and so it would be pointless to let them form. Because of this, the fungus can no longer do without the ants to achieve asexual reproduction, just as the ants cannot get enough food without the fungus. Their symbiosis has become obligatory.

This example of mutualism is so neat that one cannot help wondering whether it may not have come about through co-evolution of ant and plant, each new species of ant having as it were 'created' a species of fungus adapted to its needs. But that is in fact not the case. The basidiomycete cultivars (varieties of fungi grown by ants) found in nests are much of a muchness with ones that grow wild. This was established in 1998 in a very thorough study by Ulrich Mueller and his colleagues from the Smithsonian Tropical Research Institute (STRI) in Panama. Having searched ants' nests for fungi, they took samples of 550 cultivars; and at the same time, they gathered 300 fungi growing wild in the vicinity. They then analysed each group in an attempt to determine their 'genetic distance', a measure indicating how long ago species diverged. By drawing up this sort of genealogical tree, they hoped to be able to get back to the origin of the lineage, the very first basidiomycete fungus domesticated by the very first species of farming ant. This, however, proved to be impossible; but what they did discover was that the cultivars were very close to the wild varieties. This finding suggests that on several occasions during evolution, the ants must have renewed their stock of fungus with new wild plants, which they then redomesticated. It even appears that one ant species introduced into Florida in the twentieth century has already acquired a crop cultivated by an indigenous Florida ant species.

During their study, the team from STRI also observed that present-day ants cultivate a wide variety of fungi and that they have contrived to change their cultivars over the millennia. It has been established that distantly related ant lineages can cultivate the same fungus, while two colonies of the same species living almost next door to one another can cultivate different fungi. However, no colony ever cultivates two different varieties of fungus at the same time.

Once they have chosen 'their' fungus, the ants make a sort of contract of fidelity with it and will prevent any other variety from growing in the vicinity. This they do by using their excrement, according to Michael Poulsen and Jacobus Boomsma of the University of Copenhagen, who have conducted laboratory experiments to show that *Acromyrmex* excrete compounds which have the effect of eliminating all foreign varieties. They have even demonstrated that the ants sweep away any excreta containing another fungus, thus inhibiting the development of any vegetable competitor in their garden.

Ménage à trois

So fungus-growing ants have developed monoculture, a unique achievement in the animal kingdom if we leave aside human activity and some termites which also grow fungi in their nests. Monocultures do, however, have a drawback: they are particularly prone to parasites. But *Atta* ants have the answer to this problem: they use antibiotics.

During the 1970s, entomologists working on these ants noticed that the surface of their cuticle was covered in a white powder which they thought at first was just a secretion. However, when it was analysed by Cameron Currie and his colleagues from the Department of Botany at the University of Toronto, it was found to be a filamentous bacterium of the genus *Pseudonocardia*. There was nothing random in this. In fact, the micro-organism produces an antibiotic active against *Escovopsis*, a fungal parasite harmful to the ant cultivar. Thus, as part of their hygiene and to protect their crops, all species of *Atta* rely on the good offices of fungicidal micro-organisms. And when the queen flies away she also takes them with her and passes them on to her future family. In this way, ants, fungi, and bacteria have been getting on famously together probably for a very long time, in a successful sort of *ménage à trois*.

Part V
Bloody Pests!

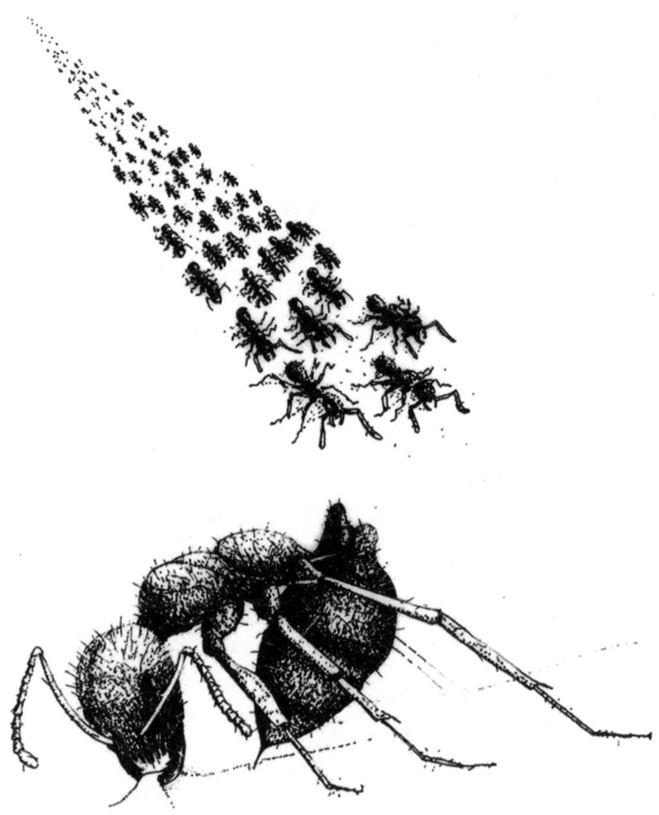

Drawing 5 Invasive ants Fire ants, taken to different continents by human activity, can overrun and devastate their new habitat.

20
Stand by for invaders!

There can be no doubt about it: in their own way, ants are geniuses. They take advantage of whatever they can use to colonize their habitat and increase their empire; if need be, they even demonstrate an astonishing sense of innovation. Just as the attines invented agriculture, *Formica lugubris* or some species of *Lasius* were into the raising of livestock long before our own era. However, despite the admiration we may feel for their behaviours and ingenuity, ants are not unmitigated paragons of virtue. They can also become nuisances, cause harm, or turn into actual pests, for instance when they invade new territories.

Most native European species, and more generally those belonging to places with temperate climates, are unobtrusive and harmless. Admittedly, they don't mind coming into your house, invading kitchens or rummaging about in rubbish bins. It is especially in spring and summer that we can see these processions of little black *Lasius niger* or their cousins *Lasius emarginatus* with their tan thorax. If they find too few aphids to exploit, they will look inside houses for the sweetstuffs they need for their brood.

European red ants, as they are known, are more aggressive and can sting. However, these *Myrmicinae*, which inhabit fields and forests, will sting only if attacked or if their nest is disturbed. Generally speaking, they are insectivores, they live in small colonies and are not invasive.

Cattle in danger

In the tropics, things are very different. Some species, such as army ants in Africa and Latin America, can become real nuisances. Nomadic as they are, they move from place to place in great hordes of tens of thousands; and when they come close to villages, the people who live there are well advised to deter them by pouring oil on the ground all round their houses. Any who do not take this precaution quickly enough find it expedient to leave home. And they had better take all their household animals with them: a cow left tethered could be nibbled away to nothing in a few hours by the strong mandibles of these voracious eaters. Against that, there is the fact that army ants do not hang about; and once they have moved on, the villagers can return home, where they will find at least that they have been rid of every last parasite. This may appear small compensation, though, for the upheaval.

Fungus-growing *Atta* ants are also near the top of the list of harmful insects. They have got themselves a bad name especially for the damage they cause to growing crops—after all, they are not called leaf-cutter ants for nothing—since for the benefit of their fungus, they wreak havoc among the fresh leaves of plants. In some tropical countries, their consumption of vegetation outstrips that of any other animal species, not excluding mammals. Nor do they bother to discriminate between wild plants and crops: if it's young and green, they take it. They are in fact a genuine pest, particularly in Latin America where *Atta cephalotes* and *Atta sexdens* can destroy up to 10 per cent of crops, an achievement which makes them champions in devastation. All told, losses attributable to attines probably amount to billions of dollars' worth per year.

Can these eating machines be eradicated? In most of the countries affected, it is seen as futile to attack them directly, as they are far too numerous and well established in their habitat for insecticides to cope with them. There is a tendency these days to starve them out by trying to destroy their favourite fungus. Research in this domain focuses on the use of fungicidal plants. However, outsmarting attines is a delicate business, since they manage to recognize such plants as bad for their crop and make a point of not introducing them into their gardens.

Invasive ants

Homegrown ants, even those that cause damage, can seem pretty harmless when compared with other species which originate from elsewhere. Known as 'invasive' ants, these are among the world's worst pests.

They often derive from South America or Africa and have been imported unintentionally into countries far away from their place of origin. These migrations were greatly helped by the explosion of international trade in plants and foodstuffs during the nineteenth and twentieth centuries, which gave them the chance to stow away in the holds or on the decks of ships. This was how they arrived on many islands, which are the ecosystems most at risk from the depredations of introduced species of any sort. But they also landed in other parts of the world, notably North America and Europe.

When they first arrive, their colonies are often small; and for some years they may pass unnoticed. However, once their bridgeheads are well established, their proliferation in the new territory is rapid. With all the arrogance of victorious armies, they set about ravaging the conquered lands. Teeming and swarming, insatiable and aggressive, they scurry roughshod over everything; the local flora and fauna bear the brunt, of course, but human inhabitants also suffer as they spread. It is hardly surprising that, of the seventeen species of invasive invertebrates seen as the most harmful on the list of the world's noxious pests, five are ants.

Entomologists have identified about 150 species of invasive ants, six of which attract most attention by their egregious ability to colonize and devastate vast tracts of territory. Take *Linepithema humile*. Originally from Argentina and Brazil, this species was first described in 1868 near Buenos Aires, which is why it is commonly known as the 'Argentine ant'. Its expansion began soon after: by 1891, it had been found in the southern states of the America and by 1904 it had reached Europe. The entomologist Luc Passera, from Toulouse, says it probably arrived in France among orchids imported from South America by horticulturists on the Riviera. From there, travelling westwards during the 1960s, hidden among container-loads of plants being sent to enhance the newly developed resorts of Port-Leucate and La Grande-Motte, the ants reached the shores of Languedoc, where they proliferated. In the mean time, they had reached South Africa in 1908 and Australia in 1939; then they arrived in Polynesia as a consequence of troop movements during the Second World War. They took longer to reach Asia, but by 1993 were established in Japan. In just under a century, Argentine ants had colonized all parts of the globe with Mediterranean climates.

Just as harmful and even more dangerous are *Solenopsis invicta* fire ants. Up to now, they have spread out less extensively because they prefer hot climates. Starting from Brazil, they reached the port of Mobile (Alabama) by the early 1920s and in a very short space of time had taken over the whole of the South. Their unstoppable advance continued: in 1996 they turned up in

Queensland in northern Australia where, within five years, despite drastic measures taken by local authorities, more than 400 sites containing millions of workers had been detected. Computer modelling suggests that, if this infestation is not eradicated, fire ants could invade between 600,000 and four million square kilometres by the year 2035. The marauding species has already reached Taiwan and China; and authorities in Hawaii are so concerned by the possibility that it might soon make landfall there too that they have already put in place an early warning and prevention programme. But it will probably take more than that to stem the advance of *Solenopsis invicta*.

Authorities in Hawaii have good reason to fear the arrival of fire ants, having already some experience of exotic pests. The island has been invaded by two of them, first the large-headed South African ant *Pheidole megacephala* and then the little fire ant *Wasmannia auropunctata*, first detected in 1999. This emigrant from South America has also managed to settle in most tropical regions of the earth. It is to be found in the Americas, from the north of Mexico to the south of Argentina, in California and Florida. It is well established on islands in the Caribbean and the Pacific, notably the Galapagos where it landed during the second half of the twentieth century and where it is advancing at a rate of 170 metres a year or 500 metres in years when conditions of temperature and humidity are optimal, for example during episodes of the climate phenomenon known as El Niño. It has also been reported in Africa and Australia, and it is on the march in Polynesia.

There is also the long-legged *Anoplolepis gracilipes*, commonly known as the yellow crazy ant and associated with human-modified environments, such as agricultural areas or urban zones. In the mid-1990s, this ant of Indian or African origin started infesting some tropical islands, including the Seychelles and Christmas Island, where it has reached high densities, devastating native invertebrate and vertebrate populations, especially threatened birds. Assisted by human activities, the pharaoh ant (*Monomorium pharaonis*) and the ghost ant (*Tapinoma melanocephalum*) have also spread to many different countries, where they become household pests, disturb greenhouse environments, and can transport pathogenic microbes in hospitals. In addition, the crazy ant (*Paratrechina longicornis*) is to be found in many tropical cities. In other words, no continent is safe from these so-called 'tramp' species: once they gain a toehold in a new country, there is no stopping them.

Nest robbers

Invasive ants leave behind not only their country of origin, but also their competitors, their enemies, and their usual parasites. Because of this, they can proliferate practically without check, at the expense of other animal species, especially the local ants which are often their very first victims. With the advantage of numbers, they immediately attack the locals or invade their nests. Being faster and better at finding food, they soon monopolize the available resources. This is too much for the indigenous ants, 90 per cent of which are estimated to have disappeared from some of the regions affected.

By looting eggs, larvae, and pupae from the nests of other insects, the newcomers disturb the life cycles of many species of local fauna. Since its arrival about a decade ago on Christmas Island in the Indian Ocean, *Anoplolepis gracilipes* has ravaged a population of twenty million red crabs. In southern states of America, fire ants stand accused of perturbing the egg-laying of several species of birds by preventing them from building their nests properly. They even attack small mammals, such as the pygmy mouse; their impact on these can be so great that, six months after fire ants have been locally eliminated, scientists have reported a 50 per cent increase in the population of the tiny mice.

Many other species suffer indirectly from the arrival of invasive ants. The Californian horned lizard, for instance, once fed on the local ants; but since Argentine ants have come along and cleaned them out, there is nothing for it to eat. Nor can it adapt to eating the newcomers, which are too small and too aggressive to be to its taste. The giant turtles of the Galapagos have other worries: they are attacked by the little fire ant *Wasmannia auropunctata* which stings them in the eyes and in their genitals, very likely causing the blindness and sterility from which they suffer. It may well be these same insects that attack the eyes of elephants in Gabon. They are certainly suspected of being the cause of the opaque corneas and blindness observed in some of these elephants.

Fatal stings

Invasive ants have no qualms about attacking creatures much larger than themselves and can even be a danger to humans. This is especially the case with the red imported fire ants *Solenopsis invicta*, renowned for their aggressiveness. Workers of this species were no doubt in the habit of defending themselves against Latin American mammals and have lost none of their nastiness through being transplanted to other countries. As soon as they touch an animal or a person, they sting, causing a burning pain.

In places with a heavy infestation of *Solenopsis invicta*, it is often difficult to detect the nests and even more so the lateral tunnels they construct just under the surface of the ground. It is wise to be very watchful, so as to avoid disturbing a nest—unless one wants to be attacked by hundreds of enraged stinging ants. An American scientist reported, for instance, that having inadvertently trodden on a mound, within a few seconds he had more than 250 stings on his leg. This is perfectly plausible, for not only are the workers apt to react in great numbers, but each of them can sting several times. They are unmatched in their ability: first, they climb the legs of their victim; next, they grasp the skin with their mandibles, swell their abdomen, insert their sting, and inject the venom, a toxic substance which, like the venom of wasps and bees, is highly reactive on the skin and can also produce more serious complications. It is estimated that one person in ten is allergic to the venom; and even a basic sting, especially if repeated several times, can often require medical treatment. Some people have an especially low resistance; it has been estimated that, in infested areas, attacks by fire ants cause about 100 deaths each year.

Among the places worst affected are several southern states of America, where studies have been done to ascertain the damage that *Solenopsis invicta* has caused to public health and the economy. In South Carolina, for example, in the single year 1990, some 5,000 people consulted a doctor after being stung, of whom twenty-seven were hospitalized and one died. In Texas, the estimate of annual medical expenses attributable to imported red fire ants amounts to some $47 million.

Though Argentine ants are not as aggressive or dangerous as fire ants, they can still become a scourge for anyone living in villages infested by them. In his novella *The Argentine Ant*, Italo Calvino vividly describes how alarmed a couple are to realize they have just taken up residence in an infested house: 'We switched on the light, one lamp for two rooms: the ants were a thick line crossing the wall from the door-frame and who knows where they came from. Our hands were already covered in them and we held them up to our eyes to see what the ants looked like, flicking our wrists to keep them from crawling down our arms.' Later, out in the garden: 'I was worried. I looked closely at the column of ants coming down the trunk of the tree and realized that the silent almost invisible seething swarm of them was also all over the ground, among the weeds. I wondered how we could ever rid the house of them.'

This vermin, as Calvino puts it at another point, can actually be a real health hazard, particularly if they frequent medical premises, as was shown by a disaster that happened a few years ago in a hospital in Chile: it was invaded by Argentine ants carrying pathogenic micro-organisms which resulted in an outbreak of dangerous infections among the patients.

Agricultural losses

Not content with attacking human beings and adversely affecting the fauna of their adopted environment, invasive ants also wreak havoc among the flora. To meet their energy needs and build up calories, they require food rich in carbohydrates, which is why they go for flower nectar, pre-empting bees and thus preventing them from acting as pollinators. Argentine ants also enjoy feeding on buds, which devastates fig and orange groves in California. Also prevalent in the southern states of America, *Solenopsis invicta* fire ants destroy electrical equipment and irrigation systems. In addition, their nests are in the shape of mounds, which reduce the amount of land available for crops; and, by making the terrain they colonize uneven, they damage agricultural machinery. In India, there are other tropical fire ants, *Solenopsis geminata*, whose depredations are mainly carried out on crops of cucumbers, tomatoes, potatoes, and cotton. Wherever they are to be found, these pests are enormously harmful to agriculture. They certainly don't let the grass, or much else, grow under their feet: in Texas alone, the cost of the agricultural damage they cause is estimated to be $300 million per year, not to mention another $200 million spent in trying to stop their ravages. Extrapolated to all the southern states, this would amount to an annual cost of more than a billion dollars.

However, though they get blamed for all sorts of evils, invasive ants may not be directly responsible for all the agricultural losses attributed to them. This, at least, is the view of the American myrmecologist Edward O. Wilson, who has recently had occasion to study two great ant plagues which befell the Caribbean. The first of these, according to the Spanish Dominican missionary Bartolomé de Las Casas, happened in 1518–19 on the island of Hispaniola (then a Spanish colony and now divided between the Dominican Republic and Haiti), which was infested by ants that destroyed a substantial proportion of the crops and invaded people's houses. The second, in 1760–1770, took place in the smaller islands of the Caribbean (the Virgins, the Windwards, and the Leewards), where the same insects appeared and ravaged the canefields, reducing them, says a contemporary witness, 'to a state of the most deplorable condition'. Wilson decided there was nothing for it but to visit the scene of the crime, have a close look at the latter-day descendants of the plague ants, and compare them to the insects described in the historical records. On Hispaniola, Las Casas describes the ants as aggressive, as having a very painful sting, living in dense colonies among tree roots and in shrubs, and invading gardens and houses. Wilson took this evidence to mean that the culprits must have been tropical fire ants, *Solenopsis geminata*. Accounts of the later episode make no mention of the aggressiveness that is typical of tropical fire ants reacting to danger: 'An attack by swarms of fire ants,' says Wilson, 'is unavoidable if an intruder nears their nests, and would surely have been mentioned by anyone who had experienced it.' He concluded from this that the late eighteenth-century episode involved not fire ants but the large-headed ant *Pheidole megacephala*, whose workers are much less aggressive, though they do enter houses.

None of this actually solved the mystery, in that the leaf-cutter ants of Central and South America were thought to be the only ones that devastated plantations. Wilson's eventual conclusion was that it was not the ants that caused all the damage but the aphids, mealy bugs, and other homoptera that live in symbiosis with them: 'The Spanish, not recognizing the role of the homopterous sap-suckers in the midst of the myriad kinds of insect teeming around their crops, would understandably put the blame on the stinging ants.' So neither fire ants nor the *Pheidole* were directly to blame for the agricultural damage. They did play a role nonetheless, for it was they who were responsible for the presence of the sap-suckers, which could not have lived without them.

21
Supercolonies

Whether they are directly or indirectly responsible, invasive ants cause huge amounts of damage. It was this very aspect of their lives that drew the attention of entomologists in the first place and made them wonder about the causes of the extraordinary ability of these ants to colonize new territories.

The ecological success of Argentine ants and the other invasive species is first and foremost attributable to their mode of reproduction. Their colonies contain large numbers of queens, which produce even greater numbers of workers. In a single year, in a single ten-hectare lemon grove in Louisiana, 1,307,000 queens and two billion workers were trapped. This makes no fewer than thirteen queens and 20,000 workers to the square metre.

So the queens are not only numerous, they are also highly prolific. Instead of mating on the wing, with all the risks this entails, they mate in the comfort and safety of the nest, well hidden from possible predators. After being fertilized, they may depart in the company of workers to found new colonies a few metres away. And even if the queen should perish during this excursion, it makes little difference, as the orphaned workers will take it upon themselves to bring up the larvae, some of which they will transform into sexual individuals with the ability to ensure the continuing growth of the 'family'.

In this way, a nest with ten or a dozen queens can rapidly produce offshoots capable of developing into a large number of new nests. If left to their own devices, they all remain close to each other. But that is where human beings get into the act: all it takes is a gardener removing a load of leaves, a rubbish collector carrying bins about, someone out for a walk who decides to dig up a lavender plant to transplant at home and, without knowing it, they have played a part in the migration of the ants and the expansion of their territory.

Unicoloniality

This is not the main reason, however, for the success of Argentine ants, which lies, rather, in their ability to construct enormous communities composed of societies within societies. The various nests are the branches of a single supercolony made up of billions of individuals who all live in harmony with each other and show no aggression. This type of social organization, known to science as 'unicoloniality', is without parallel in the world of wildlife.

And yet, at home in Argentina, these little ants (they are only about two to two and a half millimetres in length) do not cause special problems. They are actually on the humble side, which is why their species was called *humile* in the first place. As Ted Case and his team from the University of California have established, in their native habitat they exhibit standard territorial behaviour. Each separate colony lives in its own nest closely guarded by workers who forbid access to any outsiders. In a supercolony, on the other hand, all territorial boundaries are abolished; ants from different nests fraternize and go freely from nest to nest. In this way a society can extend its territory over thousands of kilometres, as has happened in southern Europe: from Galicia in north-west Spain, through Oporto, Valencia, Perpignan, and Marseilles as far as the Italian Riviera, Argentine ants have established a veritable empire stretching for more than 6,000 kilometres. It would appear that by becoming immigrants they lose their aggressive reaction towards members of other colonies.

Be that as it may, the new European representatives of *Linepithema humile* have retained all their natural pugnacity towards their prey. But it would appear that emigration has made them lose the sense of smell that once enabled them to tell ant friend from ant foe. In their original habitat, workers recognize their sisters by smell; and every colony has its own 'chemical signature' made up of a particular mixture of hydrocarbon molecules. Any outsider who strays into the vicinity of a colony is instantly detected and set upon. In a supercolony, though, nothing like that happens; and the ants seem to have become unaware of or indifferent to their own smell signals.

How can such a great change have come about? Ted Case and his colleague Neil Tsutsui put forward a preliminary hypothesis, based upon their observation that Californian populations of *Linepithema humile* show less genetic diversity than those still to be found in Argentina. They attributed this difference to the fact that only a limited number of queens had come to the new territories. And it is known that any sharp fall in the numbers of a particular species leads to a significant reduction in its genetic variability. Case and Tsutsui see this 'bottleneck' phenomenon as explaining the unwonted behaviour of the immigrant ants. The hypothesis makes sense, for if we suppose that the queens who emigrated were related and hence similar in their genetic make-up, they would also have been carrying similar chemical signatures, these smells being largely genetic in composition; and of course, they would have passed these on to their offspring. The workers would have been unable to detect the very slight differences between the smells and would have desisted from fighting with each other. The logical outcome of this would have been unicoloniality.

This all makes for a very neat theory. But is it neat enough? Our team of European entomologists decided to put it to the test in a study of the populations of Argentine ants in southern Europe. In the spring of 2000, Tatiana Giraud, a research scientist from the French CNRS and the University of Paris-Sud, collected specimens of *Linepithema humile* at different points along the coasts of the Mediterranean and the Atlantic. She hunted for the insects from the Gulf of Genoa in Italy to the shores of Cantabria in the north of Spain, finding them behind rubbish bins and on the footpaths of public parks and bringing some of them back to the University of Lausanne. In the laboratory, we set up a little arena in which we arranged jousts between workers from different regions. These tests of their aggression established that the ants belonged in fact to two quite separate supercolonies, one of them extending all over the 6,000 kilometres of the area under study, but the other one being restricted to Catalonia. Within each of them, tolerance was total; but between the two supercolonies, conflict was fierce and workers fought to the death.

Genetic cleansing

What had to be done next was determine the genetic diversity of the two populations, then compare it to that of the Argentinian stay-at-homes. We did this with Jes Pedersen, who is now at the University of Copenhagen. This led to the finding that in the two European supercolonies there was almost as much genetic diversity as there was in nests in Argentina. The 'bottleneck' theory of the Californian team was thus invalidated.

The hypothesis we then put forward was one based on 'genetic cleansing'. It posits that natural selection worked on the European immigrants in a way that favoured the fixation of common alleles affecting the workers' smell. The genes that produce chemical signatures, just like all other genes, can have various forms, called 'alleles'. Let us suppose that one particular allele is more frequent than others in an ant population, present, say, in the members of three colonies out of every four. Workers belonging to these three colonies would all share the same smell and would not fight with each other. But all three of them would attack intruders belonging to the fourth colony, which would be unable to withstand the onslaught. This would make for selection pressure in favour of transmission of the 'frequent' allele and individuals belonging to subsequent generations would eventually become less aggressive towards each other. The outcome of this would be unicoloniality.

It is quite possible that a mechanism of this type was set in motion when Argentine ants started to colonize their new environment. Those that emigrated left behind them the other local ants, their competitors. Unhindered by competition and faced now with only a restricted number of alien colonies close to their own nest, they were able to disarm, as it were. The demobilized workers had more time to devote to raising the brood and supplying the colonies. As peace breeds plenty, the colonies developed very rapidly. Seen from this point of view, it is logical that natural selection should have favoured the growth of non-aggressive genes, which ultimately fostered the spread of the species.

For the time being, this is nothing but a hypothesis which, like that proposed earlier by Case and Tsutsui, could at any moment be invalidated. The fact is that invasive species have been studied for the most part in the places they have newly colonized, which is quite logical, given that these are the places where they are causing such havoc. But relatively little is known about the lives they lead in their native environments.

Recently, working with Tatiana Giraud and Jes Pedersen, we undertook a study of Argentine ants in their countries of origin. This showed that they too form supercolonies, living in small clusters of nests untroubled by any aggression. These supercolonies, mind you, are not large, extending over only a few tens or hundreds of metres, on quite a different scale from the vast societies to be found in the United States and Europe. However, that difference in scale appears to be the only difference, suggesting that, when the ants set foot on other continents, this did not lead to a fundamental and qualitative change in their way of life. If further studies should confirm these findings, some of the theories developed to explain the behaviour of invasive ants would have to be revised.

Impossible to stamp out

One thing, however, is certain: once harmful insects are well established in a new territory, it is impossible to eliminate them. Yet attempts to do so have been made. In America, millions of dollars have been spent dumping tons of insecticides on the most heavily infested areas. But to no effect. Edward O. Wilson has used the expression 'an entomological Vietnam' to describe the nature of this lost war. The pests prevail, the imported red fire ants thus vindicating their scientific name of *invicta*, meaning 'invincible'.

Chemical warfare might prove to be less effective than a form of biological warfare using the ants' natural enemies and parasites. For instance, it has been suggested that phorid flies might do the trick: it is their curious custom to lay their eggs in the heads of fire ants, where they will develop and eventually kill their host. Such a method would certainly not eliminate the pest—for when the ants sense the presence of these flies they make fewer sorties from the nest—nevertheless, it could at least limit the extent of their invasion and enable other species of insects to regain a foothold in the affected territory. Before recruiting armies of flies, however, one would have to be sure that their parasitism targets only fire ants and not some other local species as well. Biological warfare must be waged with caution, so as to avoid introducing solutions that turn out to be worse than the problem.

Another possibility could be to encourage invasive ants to kill one another. This may appear to be a far-fetched notion; but it has been quite seriously proposed by several entomologists. Neil Tsutsui for one has suggested chemical or genetic tinkering as a way of fostering aggressive behaviour in the workers of super-colonies. He has no illusions about this possible method and is the first to stress that it would have to be subject to 'careful consideration', for it could be risky.

A better idea might be to try preventing the problems from arising, with 'a framework for identifying potential invaders', as Case and Tsutsui put it. They do add, however, that any new species discovered should be seen as 'guilty until proven innocent'. In this way, California managed over a period of some decades to block the advance of the first colonies of fire ants that were about to invade the state, by drastic measures such as the systematic checking of every single lorry entering its territory—despite which, colonies still established themselves on the west coast of America, where they have grown so large that vigilance is no longer enough. It has now become impossible to get rid of them.

Nonetheless, a solution must be found, for the unstoppable invaders keep on increasing their planetary expansion, not only throughout the tropics but also across Europe. Argentine ants being now well established along the Mediterranean coastline, another species, even more bothersome, is being viewed with foreboding by several other European countries—the invasive garden ant, *Lasius neglectus*. Originating in the steppes of Central Asia, it was first detected in 1974 in the suburbs of Budapest; and since then it has been found in many other places in Europe, such as Greece, Turkey, Romania, Spain, Germany, and, more recently, Belgium. Unlike Argentine ants, garden ants tolerate inclement winter weather. As cold cannot be relied on to halt its advance, it is expected to turn up before long in Denmark and southern Sweden, without anyone or anything being likely to arrest its advance.

All this represents the other side of a coin. Because of the phenomenal benefits confered by social life, ants have contrived to colonize all parts of the world. For these same reasons, however, harmful invasive ants have also contrived to become established in new environments. They are now real pests against which, it must be admitted, human beings are practically powerless.

Part VI
Kith and Kin

Drawing 6 Abnormal reproduction Queens of *Pogonomyrmex* mate with males of two different lineages. This is one of many ways ants can reproduce.

22
Genetic altruism and sociality

Queens are for laying and reproducing. Workers are for being sterile and busy, and for looking after the supply and protection of the colony. There could hardly be a better division of labour. It is an arrangement that has stood the test of time and promoted the spread of ants.

On closer inspection, however, it appears to be at serious variance with the Darwinian theory of evolution through natural selection. As explained by Darwin, natural selection should favour traits and behaviour increasing the chances of survival and reproduction. Yet here we have the ants, a manifest ecological success, which appear to infringe the Darwinian golden rule, since their colonies, with extremely few exceptions, are composed of vast majorities of sterile individuals. Not only that, but workers have developed over time a morphology that actually prevents them from reproducing. This looks like a conundrum. However, no Darwinian need despair: the great edifice of natural selection is not going to come tumbling down, undermined by the humble ant. On the contrary, ants actually support and reinforce it. The pains taken by the worker-daughters to look after the queen and her descendants are for the ultimate benefit of their mother as a producer of abundant offspring. In so doing, in practising 'reproductive altruism', the workers make a great contribution to the dissemination of their own genetic material.

Darwin himself had the perspicacity to suspect that ants might be a weak link in his chain of reasoning. As early as *On the Origin of Species* (1859), he spoke of 'one special difficulty, which at first appeared to [him] insuperable, and actually fatal to [his] whole theory', namely the problem of worker ants. His immediate response to this apparent paradox was to suggest that natural selection could in fact apply to the family as well as to the individual.

Gregor Johann Mendel's deciphering of the laws of heredity in the middle of the nineteenth century, then the discovery of genes in the early years of the twentieth, led to a refinement of the theory of evolution. From the point of view of genetics, individuals do not need to produce descendants of their own. They can indirectly pass on copies of their genes to future generations, so that their 'bloodline' can go on flowing in the veins of future generations.

This principle was aptly modelled by the British biologist W. D. Hamilton in the early 1960s. The theory, now known as 'kin selection', states that individuals can hand on copies of their genes not just through any descendants they themselves may have, but also by facilitating the reproduction of their close relatives. These two modes of reproduction are (almost) one and the same, since the individuals concerned share with their mother, their siblings, or even their cousins a high proportion of genes inherited from their common ancestors.

Hamilton thought his idea through, and was able to express it in the form of a mathematical equation which to this day bears his name. 'Hamilton's rule' states that an altruistic act will be favoured if $br - c > 0$. In this context, r equals the degree of kinship, as understood by probate lawyers when apportioning an inheritance among close relatives. From the point of view of reproduction, it will be in the interest of individuals to help the relatives closest to themselves on the genealogical tree, since they share with each other a greater proportion of genes identical by descent. To be complete, the equation must also take into account the benefit (b) afforded to whoever profits from the altruistic act, that is, among ants, the queen, whose benefit can be measured by the number of descendants produced. It should also be remembered that altruistic acts have a cost (c) for whoever performs them, which in the case of ants means the workers, who will never produce offspring. Those are the three parameters of Hamilton's rule; and the equation can be expressed in the following terms: altruism only makes sense if individuals transmit more copies of their genes by remaining sterile and helping their mother to reproduce than by leaving the nest and entrusting themselves to the hazards of reproduction.

As an illustration of Hamilton's rule, the British evolutionary biologist J. B. S. Haldane imagined there to exist a very special gene, one which would make the people who carry it willing to die so as to save the lives of their close relatives. If such an altruist dies, a copy of the gene dies too; but Haldane's calculation is that the frequency of this same gene would thereby increase in the population, if the self-sacrifice of that person saved at least two siblings, four nephews or nieces, or eight cousins.

Not just social, but eusocial

Hamilton's rule also offers an explanation of why ants have achieved an extreme of communal life known to entomologists as 'eusociality'.

This term, meaning 'truly social', was coined in 1966 by an American entomologist, Suzanne W. T. Batra. She used it to describe not only any societies of bees in which the parent who founded the nest cooperates with some mature daughters but also any groups within which there is an organized 'division of labour'. The meaning of the word then evolved as different schools of entomologists adapted it to their own definition. In this book, we use the adjective 'eusocial' of groups in which individuals, though they have reached the age of reproduction, do not in fact reproduce. Instead of flying off to found their own colony, daughters of the second generation remain in the nest to raise their siblings. This behaviour, though also found in some species of wasps and bees, is universal throughout the world of ants and termites.

Sociality was not suddenly magicked into existence. It must have developed dozens of times during evolution and affected different groups. Looking no farther than the Hymenoptera, we can say with certainty that it has emerged at least ten times. Nor is it restricted to Hymenoptera and termites, though for a long time this was believed to be the case. It has recently been discovered in other invertebrates, such as Japanese aphids and Australian thrips, which are tiny insects with a tapering abdomen, usually shiny black, found on many plants. It has even been found in a small mammal, the African naked mole rat, a rodent which lives in burrows and has a social organization that is curiously like that of ants (with the exception that males play a part in the life of the colony as workers, and at times even as kings).

This does not mean there is a clear-cut division between only two types of animal behaviour, solitary living or eusociality. These two states are in fact the extremes, between which there are many intermediate possibilities. There are types of prawns, various groups of birds, and even suricates (the small mongoose of southern Africa) that have adapted to living in a colony. The young stay with their parents and look after their siblings, though this does not necessarily condemn them to lifelong sterility, as it does most worker ants. If the occasion arises, they leave the nest or burrow and go off to found their own colony, as has been observed in birds which leave the premises occupied by the family and move 300 metres away to take over a territory recently vacated by a couple belonging to the same group. The future of the offspring is in fact decided by ecological constraints. If it proves possible for them to leave and reproduce, they do so; if not, it is very much in their interest to stay at home and help the parents.

It makes sense from this point of view to see sociality as a function of environment, an unexpected consequence as it were of 'the housing crisis', having arisen in habitats which make it difficult for offspring to leave their parents or their mothers and found a family of their own.

Ancestral ants probably experienced situations of this sort which made them stay in the family nest. There being few opportunities for the worker-daughters to reproduce, they would have stopped developing unusable ovaries, which would have modified their anatomy so as to increase their suitability for working. Such a scenario can account for the differences in size and appearance observable in extant ants, for if a physiological specialization once becomes established, it soon becomes irreversible.

Ants, having developed from associations of solitary wasps, became by stages truly social, adopting a mode of organized existence that evolution appears to have fostered, in particular in the order Hymenoptera. Why? Part of the answer lies in the way of life adopted by these insects. The mother and her descendants live close to each other in a nest, which may be of a very complex form. Once such sophisticated premises have been constructed, it makes ergonomic sense to house several generations under the same roof, thereby obviating needless labour.

23
Family feuds

Any worker who helps her mother to have many fertile descendants has hit on an excellent way of ensuring that her own genetic inheritance will be passed on. Her altruism is therefore not disinterested. Nor is it neutral, as workers tend to foster the development of individuals which are genetically closest to them. And to do this, they do not hesitate to engage in fratricidal behaviour.

To understand this surprising behaviour of ants, we must go back to the laws of heredity as they obtain in the animal world. With mammals, including human beings, things are pretty straightforward. Each individual, whether male or female, has two sets of chromosomes, that is, all genes in duplicate. During reproduction, the parents transmit only one copy of each chromosome to their offspring. But as the offspring inherit one copy of each gene from the mother and the other from the father, they too, like their parents, will end up with two copies of each chromosome. This mode of heredity is known as 'diploid'. Statistically, if a father has two offspring, there is one chance in two that he will pass on to them the same set of genes; and the same goes for the mother. What this boils down to is that a brother and a sister have a 50 per cent chance of receiving identical genetic copies from each of their progenitors. This degree of genetic relatedness is 0.5.

With ants, however, things are very different and heredity is far from straightforward. Females, whether queens or workers, all hatch from fertilized eggs and are therefore diploid. Males, on the other hand, hatch from unfertilized eggs. This means they have no father and the totality of their genes comes from their mother, half of whose genes they inherit. They have only one set of chromosomes and are known as 'haploid'. In a sense, they're not all there (see Figure 2).

This difference between males and females in their genetic make-up, known as haplo-diploidy, makes for great tangles of kinship relations within a family, especially between brothers and sisters. Let us imagine the simplest scenario: a monogynous colony in which the queen mates with a single male. The daughters will always receive the same set of genes from their father. But because their mother has two copies of each gene, the daughters have only a 50 per cent chance of receiving the same copy of a given gene. Two sisters have therefore the paternally inherited half of their genome which is identical and the maternally inherited half which is 50 per cent identical. In other words, their degree of genetic relatedness will be 0.75. Between females and males, the genetic similarity is lower. As with sisters, males and females have a 50 per cent chance of receiving the same maternal genes. But because males have no paternal genes, females always possess a half of their genome (the paternal side) which is wholly absent from their brothers. So, taken as a whole, the relatedness of a female to her brother is only 0.25.

In genetic terms, therefore, a worker is three times closer to her sisters than to her brothers. As a result, it was established in 1976 by Robert Trivers and Hope Hare, then at Harvard University, that workers have a vested interest in altering the proportion of males and new queens produced by their colony. This is because new queens are three times more likely to carry identical

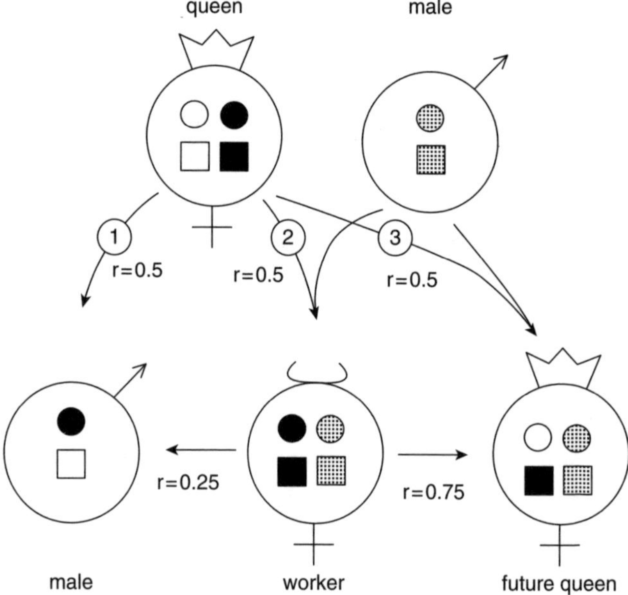

Figure 2 Haplo-diploid heredity The squares and the smaller circles represent genes and the ways they are transmitted from parents to offspring. The diagram is much oversimplified, as in reality the genome of an individual ant is composed of thousands of genes. It also illustrates the simplest case, in which a female mates with only one male. Male ants come from unfertilized eggs and so have no father. They inherit only half the genetic material of their mother (1) and have half as many genes as females. They are called 'haploid'. Females, whether workers (2) or queens (3), are the product of fertilized eggs. They have half their mother's genes and all their father's (in grey in the diagram); and they are called 'diploid'.

This haplo-diploid heredity affects the degrees of kinship (r) between individuals and makes for asymmetry between siblings. A queen shares half her genes with both her daughters and her sons (r = 0.5). Two daughters share half their mother's genes and have in common all the genes of their father (r = 0.75). But a sister and brother share only half their mother's genes (r = 0.25). Thus, genetically, workers are three times more closely related to their sisters than to their brothers.

copies of the workers' genes than the males. Hence, it is in the interest of workers that the colony should produce three times as many 'girls' as 'boys'. The equilibrium is reached at this female-biased sex ratio because when all colonies produce three times more queens than workers, males will be less common in the population and have three times more chances of finding a mate than the virgin queens. The situation is reminiscent of a punter having to choose between two lotteries, one of which promises a big jackpot while the other offers odds that are three times better but a prize that is three times smaller. For workers seeking to transmit maximum copies of their genes, the point of equilibrium is arrived at when the colony produces three times as many young queens as it does males, in other words when the sex ratio is three to one in favour of sexual females.

Mothers against daughters

Workers may have an interest in favouring their sisters over their brothers, but the queen doesn't see things quite like that. Being their mother and standing in exactly the same degree of kinship (0.5) towards her sons as towards her daughters, she has no reason to favour either of them. So mothers and daughters have divergent interests; and this underlies many conflicts within the colony.

On the face of things, it might be thought that the queen has complete power. She it is who, by opening (or not opening) her spermatheca, 'chooses' to fertilize (or not fertilize) her eggs. In this way, she can determine the sex of her direct descendants. However, the workers can also affect the outcome, in that they are the ones who tend the brood and who can therefore alter the sex ratio of the colony. They thus have the power to eliminate a proportion of either the males or the females during their development, by giving them insufficient food and care or by simply killing them.

So, ultimately, which of them, the queen or the workers, wins this competition? Following the work of Trivers and Hare, a great many teams of myrmecologists have been busy calculating the sex ratio in ants' nests, closely studying sixty to seventy different species. Their results show that neither the queens nor the workers win outright. They each have a part to play in regulating the sex ratio of the brood, though there are some species in which the mothers have the edge on the daughters and others where it is the other way round.

A recent twist to the story has come from the realization that the behaviour of workers might be affected by intraspecific variation in colony kin structure. Thus, in monogynous species, kinship relations within the colony can be greatly affected, depending on whether a queen has mated with one or more partners. If the queen mated with two males, the nests contain not only full sisters but also half-sisters, which changes everything, making it necessary to revise all our carefully calculated degrees of kinship. Let us recapitulate: half-sisters share half of their maternal genes, but as they derive from different fathers, they do not share any paternal genes. So their kinship equates to 0.25. With full sisters, on the other hand, given that they will share three-quarters of their genetic inheritance, nothing is changed. Nor is anything changed with brothers and sisters who share only a quarter of their genes, whether the queen is singly- or multiply-mated. Faced with such a conundrum, what will the workers do? How will they bias the sex ratio?

Two entomologists, Jacobus Boomsma and Alan Grafen, both then at Oxford, made the mathematical calculation. Their conclusion was that, if workers wanted to discriminate in favour of the individuals most closely related to them, they should foster females in cases where their mother queen had mated only once and males when their mother had mated several times.

This hypothesis then needed to be put to the test in the field, with real ants. Two members of our team, Michel Chapuisat and

Lotta Sundström from Finland, who was working at Lausanne at the time, undertook this task, studying *Formica exsecta*, a species commonly found both in the Swiss Jura and in Finland. They started by doing paternity tests on workers from different colonies, exactly like the tests done on people, to determine whether they had the same father or not. Then they calculated the sex ratio for each nest. And their findings largely confirmed the Oxford scientists' hypothesis. In fact, in societies of *Formica exsecta* whose queen had mated several times, the workers made no change to the sex ratio. This is logical, as they are just as closely related to their brothers as to their sisters. But when their mother had mated only once, the daughters had no scruples about killing off a great many of the males, whom they then fed to the new females.

Sniffing out the brothers

This behaviour implies that workers can distinguish between their male and female siblings. How can they do this? They certainly appear to be unable to recognize the sex of eggs. However, when males grow into larvae or pupae, it becomes easier to identify them: they are larger than females and their morphologies are different. At that stage, workers can tell them apart by sight. Or perhaps by smell. To tell the truth, we are unsure how they do it, but they can make the distinction. Evidence for this is supplied by *Formica exsecta*, a species in which, though the queen lays equal numbers of eggs of both sexes, by the time they have become pupae, there is a large majority of females. This shows that the workers have intervened during the development of the brood to kill off the males and alter the sex ratio to their own advantage.

A very different problem is how the workers decide whether to eliminate their brothers or not. In other words, how do they work out how often their mother mated? They weren't playing gooseberry; they weren't even thought of during the mating flight. The answer lies in their sense of smell. There is a genetic component in the smell of insects, as has been shown by studies made of bees and several species of ant. When they sniff their nestmates, workers must be able to detect the diversity of odours, which depends on the number of times their mother mated. In this indirect way, they are aware of the genetic diversity within their colony: if it is extensive, it means the queen has had various mates and the workers will not alter the sex ratio, thus protecting the males from their sisters' mandibles.

Whenever fratricidal conflicts do emerge within such matriarchal societies, the males always come off second best, a plight which might make one feel sorry for them. But they have a trick that can help them avoid the fate that their sisters have in store for them: they survive by 'hiding' their sex. Over the course of evolution, they have developed a mode of disguise, which consists either of smelling like workers or queens, or of being the same size as the females. This idea may sound far-fetched; but observation of parasitical ants has shown that it is perfectly possible. Once these gate-crashers have established themselves in another colony's nest, their queen gives birth to sexual offspring even at times of the year when the resident monarch is only producing workers. This means that if the intruders want to avoid being evicted, they must not draw attention to themselves. This they do, male and female, by growing to a size which is similar to that of the workers whose nest they have occupied. This has been suggested by Peter Nonacs of the University of Los Angeles; and we have confirmed his findings in studies of the tiny *Plagiolepis xene* ants in the south of France. Among this parasite, sexual individuals, whether male or female, are all of a size with the workers which they exploit. If they grew any larger, they would be detected and eliminated; and if they were any smaller, they would have less chance of ever founding new colonies. So natural selection sometimes works in mysterious ways; and it is highly possible that some males have benefited from this to escape their sisters' clutch.

Males that do not resort to this survival ploy may be able to count on the protection of their mother. For queens also know a trick or two. They can, for instance, give the workers less room for manœuvre by varying the number of males and females in the eggs they lay. Red imported fire ants do this, as demonstrated by Luc Passera and Serge Aron working with our own team as well as with Edward L. Vargo, then at the University of Austin (Texas). There are colonies in which the queen gives herself complete control of the situation by producing almost exclusively male eggs, which means that since there are not enough females, the workers have no choice but to tend to the males—a second best no doubt from their point of view, but better than nothing.

It appears that, in general, queens have greater control over colony sex ratio in monogynous than in polygynous nests. When there are several queens in a colony, they compete with each other. Even if most of them produced a majority of males, it would require only one or two of the others to produce females for the workers to favour them. When there are conflicts within a dominant class, the dominated can take advantage of the situation to promote their own interests. To this general rule of societies ants are no exception.

24
Nepotism or not?

The presence of several queens in a colony who mate with several partners can produce a genetic imbroglio in which workers are either full sisters, half-sisters, or cousins, or even possibly quite unrelated to any other individuals. In other words, every type of reconstituted family is possible!

On the face of it, the rules governing kin selection should mean, strictly speaking, that it would be very much in the interest of a worker to favour the most closely related nestmates over less related individuals. In other words, one might expect that colonies would be subject to governance by nepotism. Is this, however, the case? Up to the late 1980s, entomologists believed so, on the strength of laboratory studies made on honey bees whose queens had mated with several males. Observation of the behaviour of workers had led to the conclusion that they did in fact adhere to the rule of nepotism by favouring the production of queens belonging to their own paternal lineage.

These studies have more recently attracted criticism, first and foremost because they did not replicate real situations. In the bee experiments, the queens had been mated with only two males, whereas in the wild they have a great many more partners. In addition, the various lineages of workers, born from different fathers, behaved differently, some being, for instance, more active than others, in particular in that they were better at tending the brood. So, for one thing, the findings could have been biased. The real problem, however, lay in the fact that at that time entomologists used fathers with genetic differences, which made it possible to distinguish offspring on the basis of morphological or physiological differences. These genetic differences may have affected the behavioural tests performed.

Nowadays, advances in molecular biology mean that it is no longer necessary to use males with mutations, allowing us to recognize their offspring. To determine the degree of kinship between two individuals, we now analyse small parts of their genome, known as microsatellites. These are in fact particular fractions of DNA, pages of the great book of life of the genome, on which certain 'words' are repeated many times, the number of typographical repetitions varying from one individual to another. As the microsatellites of any offspring are a combination of those of the parents, it is possible to test them for paternity or maternity.

Entomologists had to start again from scratch, using the genetic tests in studies of the behaviour of bees, wasps, and ants, which in addition were allowed to mate under natural conditions. Apart from a few exceptions, the new studies point to the absence of nepotism in social insects.

This finding, when you think about it a bit, makes complete sense. If all the workers set about favouring only their sisters and half-sisters and eliminating all other females in the brood, the nest would live in a state of chaos. In any case, no system of recognition is infallible, and ants do make mistakes when trying to tell who is related to whom. They could get carried away and eliminate also some of their own lineage, which would add to the prevailing disarray. Such a state of affairs would be beneficial neither to individuals nor to the colony as a whole. Quite the opposite, in fact.

To avoid any tendency towards nepotism, natural selection has favoured mechanisms which reduce ants' ability to recognize others of their own close kin. This hypothesis is also consistent with the findings of studies made on bees and a species of wood ant. Chemical analyses of hydrocarbon mixtures on the cuticles of ants reveal that individuals belonging to the same lineage produce similar smells. So, each family has one odour blend that is peculiar to it, though it is still very close to those produced by other lineages. Moreover, because cohabiting in the same nest brings about a standardizing of the various smells, workers end up being unable to recognize their closest kith and kin. The whole colony functions as though it has developed a degree of social uniformity that works towards the avoidance of discrimination.

25
Caste struggles

So workers do not go in for nepotism. This does not mean, however, that they lose sight of their main aim, which is to pass on their genes to subsequent generations. They have just hit on other ways to achieve this goal: they do it through affecting the castes. In other words, depending on what suits them, they favour the development either of new queens or of workers. This makes for further conflicts among the various generations competing to defend their own interests.

The small *Leptothorax acervorum* ants keep their struggles covert. A pair of British entomologists, Robert Hammond and Andrew Bourke, have shown that the number of queens influences the degrees of kinship among nestmates and, if the nursemaids are genetically closer to the future queens than to the males, they neglect the latter and invest the greater part of their energy in feeding the former. They do not kill their brothers; but they favour the development of the reproducers-to-be, which is basically their roundabout way of altering the sex ratio of sexual individuals produced by the colony.

But this manipulation of castes has the side-effect of favouring the development of the young queens at the expense of future workers. If this bias in favour of producing an aristocracy rather than the working class should be taken too far, it could jeopardize the whole colony. It is difficult to see how any ant society could function like a 'Mexican army', with more officers than privates. Nor would the queens and their brood survive for long without the help of their regiments of workers to feed and defend them and maintain the nest. In fact, whether it is more beneficial for a colony to produce a majority of queens or of workers depends on its age and size. A colony that is old and large enough can cope with producing great numbers of new queens. It already has enough workers and a few more or less will not in practice have much effect on either its chances of survival or its productivity. This would of course not be the case with the youngest colonies, where the labour force is thinner on the ground, hence more valuable.

The larva has its say

Even so, the question remains unanswered: who allots caste among the females of the brood? Whose influence is it that will turn young females into new queens or workers? The queen's? The workers'? Or could it be by any chance that the larvae do it themselves? When we suggested, about twenty years ago, that the baby ants might have a say in the matter, colleagues greeted the idea with great scepticism. The fact is that at the time many entomologists were convinced that the queen was responsible for everything; and it is also a fact that she produces pheromones which were thought to have the effect of inhibiting the development of larvae into fully sexed females. It turned out, however, that this was not the case. In the late 1980s, we and Peter Nonacs suggested that the primary function of royal pheromones was to enable the queen to communicate with the rest of the colony; the mother saying to her daughters, 'You see, I'm in good health, I'm fertile, and if you help me, I'll give you many siblings.'

Since then, many studies have been undertaken which confirm that the queen's pheromones indicate her reproductive status and her state of health. If she lays a lot, she produces molecules in great quantities. But if she dies, her daughters stop detecting the chemical signal, which is a sign that they must devote all their energy to producing new queens.

The issue of who is in control of caste determination is of importance because the interests of the different castes of females are not necessarily the same. Some entomologists have seen this difference of interests as a potential source of conflict between queens and workers, it being to the latter's advantage to favour the development of queens rather than workers. Others took the view that this need not be a source of hostility, since queens and workers have a common interest in favouring optimal production of workers. Workers after all contribute to the development and longevity of the colony, thereby enabling it to go on producing great numbers of reproducers, whether male or female.

Working on mathematical modelling with the German scientist Max Reuter, we succeeded in demonstrating that the outcome of conflict between queens and workers depends on the numbers of diploid eggs available. If a colony has a great many female eggs, the workers can produce as many young queens as they like without compromising the numbers of their fellow workers. In a situation of plenty, both interests can be served without any conflict arising. However, when the supply of diploid eggs is limited, it may still be in the workers' interest to favour the development of queens, even though this results in fewer workers produced and thus smaller colony size. And in such circumstances, competition and hostility may arise between the queen and her workers.

The larvae, too, have their own genes to look after. A larva that lives to have descendants of its own will be more closely related to its daughters and sons than to its nieces or nephews. It is therefore very much in the interests of larvae to become queens, even though that may go against the interests of their adult nestmates. In this lies the germ of further potential rivalries, which are more likely to turn into real conflict in species where the queens and workers are rather similar in morphology and size. In such cases, the workers' room for manœuvre in allotting more or less food to the larvae, thus influencing what will become of them, is limited. When the nursemaids lose many of their prerogatives, larvae have more scope for determining their own development.

Conflicts of this sort have already been observed in polygynous nests among certain species of ants and stingless bees. When the quota of queens is filled, the remaining larvae are forced by the adults to end up as workers; and if not, they are simply eliminated.

Clearly, as all these conflict situations arise from issues of kinship, any divergences between females are a consequence of the number of mothers present in a nest and of the number of males they mated with. In theory, by calculating the degrees of kinship involved in the various scenarios, one can predict what should develop. If workers and larvae control the allotting of caste, their favouring of the production of new queens should depend on the number of partners with which the queen has mated; whereas, if it is the queen who is in charge of caste allocation, the number of times she mates should have no bearing upon the proportions of young queens and workers in the colony. So much for theory. In practice, however, only observation of the different scenarios in the field will be able to determine whether the queens, the workers, or even possibly the larvae, have the upper hand in this entomological version of the class struggle.

Problems with males

If one thing is certain about ants, it is that, apart from a very few rare species like those in the sub-family of the *Ponerinae*, workers cannot ever mate. For they have no spermatheca, the pouch that enables queens to keep a stock of males' spermatozoa throughout their lifetime. However, there is nothing stopping workers from laying eggs. Even if unfertilized they can always serve as a source of food for the colony or, if they develop, hatch into males (which it should be remembered always come from unfertilized eggs). In monogynous nests that have lost their queen, the workers actually succumb to the desire for motherhood. Because there is no more future for their colony, their best option is then to reproduce on their own.

When the queen is present, however, the workers generally remain sterile. This is especially remarkable since, according to the theory of kin selection, they would be better served by reproducing, thereby passing on more copies of their genes. If a queen has mated only once, the workers will share with their own sons half of their genetic inheritance, whereas their degrees of kinship with their brothers (who are male offspring of the queen) or their nephews (offspring of other workers) will drop respectively to 0.25 and 0.375. It is conceivable that, as they are closer to their nephews than to their brothers, it would be to their advantage to help their sisters reproduce rather than their mother. But the queen, too, has more to gain in the matter of heredity if she gives birth to her own male offspring, with whom her degree of kinship is 0.5, than if she is content to have grandsons (0.25). In this area as in others, the interests of the queen and the workers diverge, heralding new conflict between mothers and daughters.

But even this situation can change if the queen has mated with several partners. The workers are then genetically closer to their mother's sons than to those of their sisters and half-sisters.

So they are not going to let their equals reproduce. They keep each other under surveillance, functioning within the colony as what entomologists call 'police workers'. This system of monitoring works very effectively, as has been shown by studies of bees, who also do such constabulary duty. If a worker suddenly develops its ovaries, her nestmates lose no time in attacking her or in destroying any eggs she may have laid. In either case, the offender has little chance of succeeding in her designs. Myrmecologists have concluded that the setting up of such a system of policing was essentially an outcome of the kinship ties within the colony.

But this behaviour can be explained in a quite different way, because it is likely that worker reproduction incurs costs for the colony. If all the workers started procreating, they would expend a great deal of energy in laying, which would have a detrimental effect on their daily work. There would be many fewer legs and mandibles available for protecting the nest and feeding all its inhabitants, and eventually the whole colony would be worse off.

Which of these two hypotheses, the one based on kinship, the other on productivity, is the more tenable? With the aim of resolving this question, we analysed in collaboration with Rob Hammond all the data available, which meant we could compare the reproductive behaviours of workers in different colonies belonging to fifty species of bees, wasps, and ants. Our first finding was that, in 90 per cent of these species, the majority of the males are offspring of queens and not of workers. This proves that workers make little use of their ability to give birth to their own descendants. We also found that the workers who do reproduce are no more numerous in colonies where there is high genetic diversity. Even in nests where they are more closely related to their nephews than to their brothers, and where procreating would therefore be in their interest, they still produce very few males. All this supports the case for productivity being the main factor explaining why workers do not usually reproduce. In a more recent study, which adds more data, Tom Wenseleers and Francis Ratnieks confirmed our finding that worker reproduction is rare, but they also found a tendency of workers to be less likely to reproduce when the queen has mated with several partners. Thus, variation in kinship across colonies probably also plays a small role in explaining variation in the reproduction of workers.

26
Anything goes

Close observation of ants shows that even in reproductive matters they like to display originality. Some queens have no objection to mixed marriages and will mate with males from outside the family. Others enjoy having it both ways and go in for sexual and asexual reproduction. There are even some who use cloning to reproduce descendants in their own image. Their males, not to be outdone, do the same. Among ants, it seems, anything goes.

Cross-breeding

If you're a female ant, whether you lead your adult life as an aristocrat or a prole will depend on the good offices of the workers. If they give you enough to eat as a larva, you end up as a queen; if not, you join the ranks of the barren, who do all the work. Among social insects, there is nothing more normal. However, this is only a rule, and like all rules it has its exceptions. It was believed for a long time that caste was invariably an acquired characteristic. But in fact the determination of caste can at times be innate; in other words it can have a genetic origin.

Myrmecologists were first alerted to this by *Pogonomyrmex*. American entomologists recently discovered that in this species of harvester ants from New Mexico there are two types of males, different in colour, some red and others black. Genetic investigation proved to their satisfaction that the males belonged to two different lineages. Instead of finding anything untoward in this, the queens mated indiscriminately with males from either lineage. What is surprising about this is that normally in the animal kingdom reproduction requires the bringing together of two partners of the same species. What is even more surprising is that what becomes of the daughters depends on the genetic origin of their parents. If both its parents derive from the same lineage, then a fertilized egg will develop into a queen. But if it is the product of a mixed marriage, it will turn into a worker. This means that young queens produced within the colony may also, depending on the origin of their parents, belong either to the red lineage or the black. All the workers, though, being outcomes of cross-breeding between the two lineages, will be identical.

There was something in this that had to be examined more closely. So, with the American entomologist Sara Helms Cahan, who was working at the time in Lausanne, we pursued the genetic analysis of harvester ants. Our special focus was on the DNA contained in the nuclei of cells; but we were also interested in the 'mitochondrial' DNA, a particular part of the genome transmitted to offspring only by the mother. This examination led us to conclude that the genetic composition of the two lineages was in fact a hybrid (a mixture) deriving from the genomes of two different species of ants still to be found in New Mexico, which must have cross-bred at some time in the past, possibly on several occasions.

Nowadays, the two lineages of *Pogonomyrmex* do not just get on well together, they actually cannot do without each other.

The fact is, the queens have to mate with males of both colours. If not, they would either produce solely workers, which would seriously jeopardize the transmission of their genes to future generations, or else solely queens, which would mean that, without workers, the colony could not survive. These risks are avoided by the queens' mating on average with six partners, among whom there is a strong chance that there will be both red and black males.

In some species of fire ants in Texas, Helms Cahan observed something very similar to these arrangements of the New Mexico harvesters. Though there are two types of male fire ants who coexist, the queens all belong to the same lineage. Here, the young queens mate only once, but the eventual balance of castes is guaranteed by the presence of several queens in the nest. Those which mate with males belonging to their own species will have reproductive offspring; those which mate with 'foreign' males will produce workers. So, taken as a whole, the colony will continue producing both queens and workers and thus will be able to develop in a properly balanced way.

No sex, please

Once upon a time, if there was something that could be taken for granted about ants, it was that reproduction depends on mating, that is, sex. Admittedly, males come from unfertilized eggs; but against that, all females have a father from whom they receive half of their genome. However, ants are full of surprises; and though most species do in fact follow the norm, there is at least one that contradicts it. This is *Cataglyphis cursor*, whose queens behave with reckless abandon, sometimes using male sperm to have daughters and sometimes not bothering. Sexual and asexual reproduction are all the same to them.

Sexual reproduction has been a bother to biologists for a long time. What is sex for? Why should sexual reproduction be so widespread throughout the animal kingdom? If you think about it, this way of having children is rather inconvenient, given that it requires two partners of opposite sex to come together and that this is a matter of chance. From the genetic point of view, this is particularly unprofitable for the mother, since it restricts her to passing on only half of her genes. In addition, she invests half of her reproductive energy in having male offspring, who are not very useful except for fertilizing eggs. This makes for a very wasteful system, marked by what the British biologist John Maynard Smith famously defined as 'the twofold cost of sex'. Nevertheless, there is much to be said in favour of sexual reproduction. For a start, the mix of parental genes passed on to children makes for increased genetic diversity in the population. Over evolutionary time, this constant mingling of genetic inheritances leads to the elimination of deleterious genes and the preservation of functional ones. A noteworthy long-term consequence of this is that it produces individuals who are more resistant to parasites and other pathogens.

This is the reason why, although asexual reproduction has appeared many times, species that go in for it do not last long in evolutionary terms. Some species of arthropods, some insects, and some lizards still contrive to do without sex, but they are a tiny minority in the animal world. Ants, however, do not belong to that small category; and myrmecologists were for a long time convinced that all queens had recourse to sexual reproduction. Witness any mating flight and it will be apparent that queens are not averse to copulating with males. However, close study of *Cataglyphis cursor*, with Morgan Pearcy and Serge Aron of ULB and Claudie Doums from the University of Paris VI, led us to recognize that there was more to it than that. The interesting point about this particular species was that some of the behaviours of its queens are unusual. In these monogynous colonies, the young virgin queen does not fly away in search of a mate. All she does is leave the nest and wait until males come to her; at that point she copulates with them, still on the ground close to the nest, before going back inside as soon as the love-making is over. She stays in the nest only as long as it takes to gather together a party of her sister workers, then sets off with her troops to found her own nest a few metres away.

Intrigued by *Cataglyphis cursor*, we collected thirty-eight colonies in the south of France and analysed their DNA, using genetic markers, with the object of determining who had begotten whom and how. To our great surprise, we discovered that the genotype of the young queens was not the same as that of their sister workers. It eventually turned out that the vast majority, 90 per cent, of young queens were the outcome of asexual reproduction. They had not hatched from fertilized eggs but were produced via parthenogenesis. What is going on here? Remember that all female ants (queens and workers) are diploid, that is their cells contain one set of chromosomes from each parent, whereas male cells are haploid, as they receive only the maternal chromosomes. Thus daughters born via sexual reproduction inherit two sets of chromosomes, one maternal and one paternal. In the case of parthenogenesis, however, there are no paternal chromosomes. The maternal cell divides, each of its two parts bearing a set of chromosomes, which then recombine to form the daughter cell (see Figure 3). This means that young *Cataglyphis cursor* females of royal caste, having no father, inherit all their genes from their mother. Workers, on the other hand, follow the standard pattern of females and hatch from fertilized eggs.

Thus the queen has it both ways, eating her cake and having it. By producing new queens via asexual reproduction, she avoids paying the cost of sex, because she hands on all her genes to her reproductive daughters. But by using sexual reproduction to give birth to barren daughters, she manages to maintain the colony's

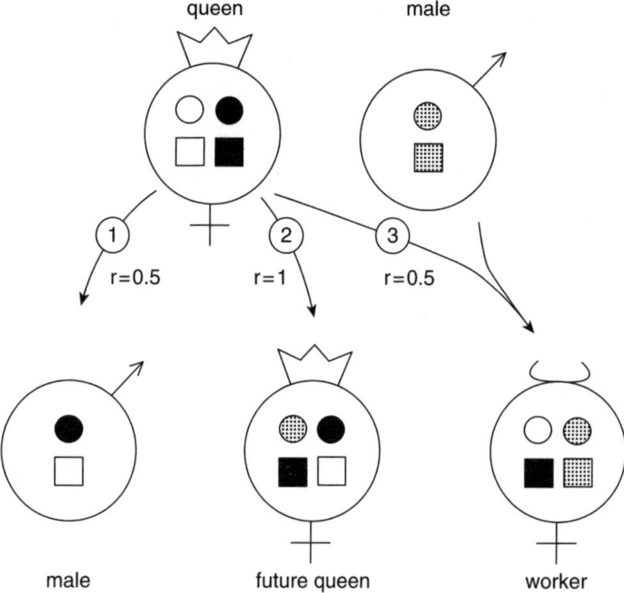

Figure 3 Parthenogenesis In *Cataglyphis cursor* ants, queens produce their sons and their worker daughters by the 'normal' method. Thus males, from unfertilized eggs, receive half the genes of their mother (1), while the workers, produced by sexual reproduction (3), inherit half their mother's genes and all their father's. On the other hand, mothers produce young queens asexually, by parthenogenesis (2). In this case, the maternal cells divide in two, each of them bearing a set of chromosomes that recombine to make a daughter cell. So the young queen inherits genes only from its mother and the two females have a degree of kinship (r) equal to 1. However, because of the chromosome recombinations that occur during parthenogenesis, the genomes of the mother and daughter are not identical.

genetic diversity and reinforce the workers' resistance to parasites. This, as the workers are in constant contact with the outside world, is very advantageous for them. If the queens, by being produced asexually, are more prone to pathogens, their risk of being infected is reduced, as they are constantly protected by the workers.

In all of this, the males are once again the losers. If they are to pass on copies of their genes to posterity, their only hope is to find one of the rare queens hatched from a fertilized egg, a thing which can happen, given that there are always exceptions to the rule and about 10 per cent of the female reproducers are not born via parthenogenesis. Even so, the males' reproduction rate is pretty low. The only way they can improve it is if the queen dies, because then the workers make haste to replace her by producing young queens. Being unmated, they can of course only do this via parthenogenesis; but that is immaterial, since they are passing on their paternal genes to the future queens, who will transmit them in their turn to their own progeny. All things considered, the males hand on very little of themselves to posterity and appear to be duped all along the line.

Double cloning

Obviously, for *Cataglyphis cursor* males, it's a mug's game: the queen wins hands down and, even if the males do on occasion manage to pass on their genes, overall their reproductive success is very poor. However, there is a species, the little fire ant, in which the queen produces her royal lineage by cloning but the male reproducers do not let her get away with exploiting them. They have hit on a smart solution: to prevent the queens from monopolizing reproduction for their sole benefit, the fathers turn out sons in their own image—that is, they too have recourse to cloning. This is unique in the animal kingdom, where such a mode of reproduction is almost unheard of; and when it does occur, it appears to be the sole preserve of females.

It was by the greatest of flukes that we discovered this utterly atypical behaviour, our real focus having been on the invasiveness of the little fire ant, *Wasmannia auropunctata*. Our object was to find out the history of introduced populations of this polygynous ant, in which new colonies arise by budding, a process whereby young queens establish their new colony not far from the nest where they were born and from which they then poach workers.

Working with Denis Fournier and Arnaud Estoup of the Institut national de la recherche agronomique (INRA) at Montferrier-sur-Lez, France, and with colleagues from the University of Toulouse III and a laboratory at the Institut de recherche pour le développement (IRD) in New Caledonia, we undertook a study of *Wasmannia auropunctata* ants collected in French Guyana. This was how we discovered that, in any given population, not only were there great genetic differences between the queens and the workers, but that all the queens had the same genotype. On closer inspection, we realized that the queens also reproduced asexually. However, unlike *Cataglyphis cursor*, they do it not by parthenogenesis but by cloning. The difference lies in the fact that in parthenogenesis the maternal alleles can recombine. The genes of the queen exist in fact in two forms, A and B, say, which means that the daughter may receive different copies, AA, BB, or AB. Because of this, mother and daughter do not have identical genomes. And there lies the difference with cloned reproduction, which always reproduces a daughter who is an exact replica of the mother.

All this was surprising enough; but the ants' genes had many more surprises in store for us. As we analysed the genomes of the workers, we realized that they had all inherited the same paternal genes. Could this possibly mean they all had the same father? That would imply that a single male had been smart and prolific enough to have contrived to mate with every single queen. This seemed too implausible to be true, even though the nests were quite close to each other. The only other possible explanation was to suppose that all the males in the species had the same genotype. To test this hypothesis, we conducted genetic analyses of all the individuals produced in colonies where we knew the

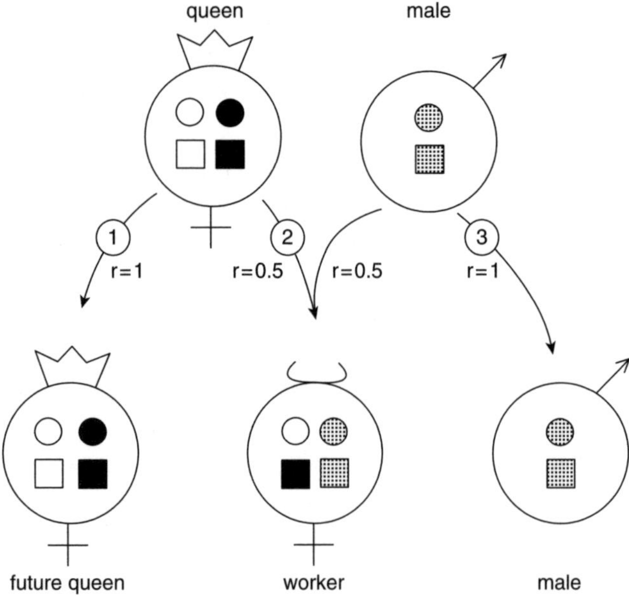

Figure 4 Cloning In the fire ant *Wasmannia auropunctata*, reproduction breaks all the rules. Only workers come from 'normal' sexual mating (2). Their degree of kinship (r) with their mother is therefore equal to 0.5.

On the other hand, mothers produce young queens by cloning (1); that is to say, they transmit to them the entirety of their genome. So queens, both mothers and daughters, are exact genetic copies, with a degree of kinship equal to 1.

Males also reproduce by cloning, producing sons with the same genome as themselves (3), so their degree of kinship is also equal to 1.

Because of this dual cloning, the only one of its kind known in the whole animal kingdom, *Wasmannia auropunctata* queens and males are not related. It is as though they each belong to different species.

genotype of the queen and her mate. These analyses unambiguously confirmed that only the workers are produced by 'normal' sexual reproduction. All the new queens were genetically identical to their mothers, while young males were similar to their fathers (see Figure 4).

This can only be possible if the males are also produced by fathers cloning themselves. This makes all the males true twins. On the face of it, this seems not easy as it is the queen who is laying the eggs. However, nature has found the knack of helping the males to perform this little conjuring trick. At fertilization, their spermatozoa, which are kept in the queen's spermatheca, penetrate the eggs and reach their nuclei. Once they have managed this, some of them apparently eliminate the maternal genome and take its place. Is this not a fine example of genetic identity theft?

So, in the little fire ant, there is no intermingling of reproductive castes: queens produce queens, males produce males, and there is not the slightest possibility of mixing bloodlines. This way of reproducing is not without consequences, for it makes impossible any gene flow, that is mixing of genes, between males and queens. The only individuals who could have contributed to the genetic diversity of the population are the workers, for they have inherited chromosomes from both parents. Given, though, that they are sterile and have no progeny, there is no role for them to play. The upshot is that the queens and the males behave as though they belonged to two separate species.

This mode of reproduction is absolutely without parallel anywhere in the animal kingdom. Of course, some species of fish, amphibians, and insects reproduce by natural cloning; but it's always the females among them who do this, by eliminating the paternal genome. It never happens the other way round. So the little fire ant has come up with an innovation and has taken asexual reproduction one step farther. It is likely that during evolution these ants behaved at first as *Cataglyphis cursor* still do, the queen mother producing young queens by parthenogenesis. But then the males, almost deprived of any possibility of reproducing, got their own back by devising a daring strategy for handing on their genes, for all the world as though they were saying to their partners, 'You think you can pass on your total genome intact to your daughters? Well, in that case, we'll do likewise with our sons!'

Could this be an ingenious solution to the recurring problem of sex rivalry? Be that as it may, the little fire ant *Wasmannia auropunctata* certainly gives us a new glimpse of the richness and complexity of methods of reproduction among social insects.

Part VII
Sociogenetics

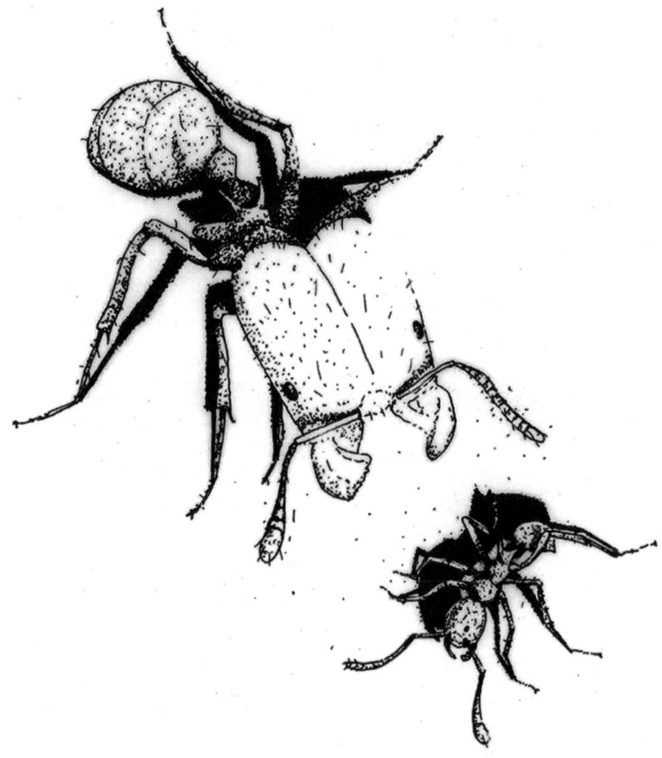

Drawing 7 One genome, two phenotypes *Pheidole* ants have the same genome, but their genes, depending on how they are expressed, can produce different castes.

27
Genes and family structure

There is nothing like genetics for finding out about the private lives of ants and discovering the various modes of reproduction that different species have adopted. But advances in molecular biology can take us even farther into their secrets, going beyond the general to the particular. Nowadays it is possible to bring to light subtle variations between the genomes of individual ants within the same species and to link genetics to behaviour.

Two questions exercise the minds of specialists in the life sciences today. Does behaviour have a genetic basis? If so, how can the genes be identified? Hitherto, in their attempts to find answers to these questions, biologists have relied on their favourite laboratory animals: mice, the fruit fly *Drosophila*, or even the worm *Caenorhabditis elegans*. But these solitary organisms are not very helpful in studying genes involved in social behaviour. Recent studies of social insects have produced findings that are much more promising. Work on the red imported fire ant has actually identified a genetic element which does directly and indisputably influence the social organization of colonies—a world first.

The fire ant *Solenopsis invicta* really is very special, for it can have two types of organization, the monogynous and the polygynous social forms. In addition to the difference in numbers of queens per colony, there are other differences between the two types of organization. A young queen striking out on her own from a monogynous society does not hesitate to fly up to several hundred metres above the ground in search of a suitable mate, before going on elsewhere to found her family, whereas queens from polygynous societies fly neither so far nor as high, and once copulation has taken place, they either go back to their original nest or take up residence in some other polygynous nest, where they set about reproducing.

Thus within a single species there can be two quite different types of family structure associated with marked behavioural differences. This we found strange and intriguing and it made us wonder whether there was gene flow between the two modes of social organization. In other words, was it possible for cross-breeding to happen between individual ants belonging to a monogynous nest and a polygynous one? Or did the queens' relationships never change, those from monogynous nests mating only with males from the same background as themselves and the polygynous ones doing the same?

In a collaborative study with Kenneth Ross of the University of Georgia, we discovered that in fact the males were not all that selective. In particular, many of those that had been born in single-queen colonies mated with females from polygynous nests. This is the reason why the genomes of individuals from the two types of nests are, broadly speaking, very similar. There is, however, one exception to this, and it is a highly significant one. We discovered that there was a particular gene which made it possible to differentiate individuals belonging to monogynous colonies from the others. This gene is called 'Gp-9', meaning 'General protein-9'.

General protein-9 can have two forms, two alleles, designated B and b. Given that females have two copies of this gene, this gives three possible combinations: the genotype can be either BB or bb, individuals having these being called 'homozygote', or else it may be Bb, in which case the term used is 'heterozygote'. Populations should therefore contain a mixture of these three genotypes. But we found that in monogynous colonies all queens and workers were homozygotes and all were BB. In polygynous nests, however, all the queens were heterozygotes (Bb), while the workers were either BB or Bb.

This finding was amazing, because it contradicted Mendel's genetic laws: according to the principles of heredity drawn up by 'the pea plant man', there must be something wrong with it. In polygynous colonies, we should also have found BB and bb queens as well as bb workers. But there weren't any.

The characteristics of the alleles offer one possible explanation for this. To have two copies of b present is lethal, and any queen or worker carrying them do not live long after reaching adulthood, perhaps some weeks or even just a few days. So there is nothing surprising in there being no bb queens to be found in the nests, since they all die young.

But why are there no BB queens in polygynous colonies? Was there perhaps some morphological or other peculiarity which might be working against them? We decided to investigate whether there was any link between the genotype of the queens and their phenotype. And we found that in fact there was. It turned out that BB queens were the heaviest and most fertile of all and that bb queens (we did actually find a few young ones still alive) were the lightest and least fertile. The Bb queens were somewhere in between.

All this was greatly perplexing. If BB queens are so fertile, why are they absent from polygynous nests? We did in fact find some, only among the young unmated queens, but this does not really alter the problem, for in theory they should have been very much in the majority. After much pondering, it occurred to us that the b allele might turn the Gp-9 into a 'selfish gene'. This is the term used for a gene that favours its own propagation to the detriment of the individual carrying it. Such a gene in *Solenopsis invicta* might make heterozygotic Bb workers eliminate BB queens. This is prima facie possible, given that after copulation queens from polygynous nests make themselves at home in an existing nest. It would be possible for the workers who live there to sort out the sheep from the goats, accepting some of them and violently reacting against others.

To put this hypothesis to the test, a simple experiment was required. To see what would happen, we put queens with different genotypes into polygynous nests, or rather, the queens we used were randomly chosen, and only after we had observed their behaviour did we analyse their genome. The results were not long in coming: all heterozygotic queens, that is to say those who had the b allele, were accepted by the colony. The BB queens, however, did not last long: less than a minute after being put into the colony, they were killed by the workers. Most of the workers prominent in the attack were heterozygotic. This was enough to demonstrate that the b allele does play an important part in this whole matter: it is the reason there are no BB queens in polygynous nests.

To call such a gene 'selfish' seems very appropriate, for its action favours its own propagation to the detriment of other variants of the gene. It would follow from this that the incidence of the b allele should increase in the population. On the other hand, given that it is lethal, it cannot achieve complete dominance. Selection is therefore subject to two opposite forces which ultimately cancel each other out, thus maintaining a state of equilibrium between the B and b alleles in populations of *Solenopsis invicta*.

Driven by their genetic selfishness, the workers enforce their law. This leads immediately, of course, to other questions. How do they manage to distinguish the acceptable queens from the unacceptable? And why is it that workers which had taken part in the killings were themselves often seen to be later eliminated by their own nestmates? The answers to these questions, as to many others relating to ants, are to be found in smell. This too we were able to demonstrate experimentally, by rubbing workers against the cuticles of different queens before putting them back into their own nests. Those that had been in contact with Bb queens had no trouble being readmitted, but those that had acquired the smell of BB queens were not so lucky: they were attacked and most of them were killed by their nestmates.

It is likely that the Gp-9 protein, 'made' by the gene of the same name, plays a role in influencing the smell of ants. In 2002, Michael Krieger and Kenneth Ross isolated the protein and analysed the DNA sequence of the corresponding gene. Comparison of the sequence with others already stored and classified in a gene bank showed that it had a close similarity to an 'odorant-binding protein' already discovered in moths. In fact, the molecules attach to the pheromones which they carry into the insects' bloodstream, thereby possibly influencing the scent of the queen. Interestingly, genetic variations associated with the Gp-9 gene play an analogous role in three other closely related species of fire ants. What we have here therefore is the first genetic element ever to be identified as influencing social organization in any living creature.

Social environment

Not that genetics is the only factor influencing these types of behaviour. The social environment in which individual ants live also plays a part in the colony social organisation and worker behaviour. Amazing though it may seem, a change in the numbers of heterozygotic workers can be enough to turn a colony of monogynes into a colony of polygynes, and vice versa.

If a colony is composed of heterozygotic workers it is polygynous. Also, it has no objection to the arrival of a great many queens, on condition that they too are Bb. If, however, it is a society composed solely of homozygotic BB workers, it will be monogynous and will allow no other queen of any sort to set foot inside the nest.

This behaviour is paradoxical, since the same BB workers, as long as they live in polygynous colonies rubbing shoulders with their Bb nestmates, can be entirely hospitable to an alien queen. It is this fact which made us suspect that their behaviour is influenced by their social environment.

To put this idea to the test, we decided to change the proportions of BB and Bb workers in a number of nests. We took a Bb queen from a polygynous colony and tried putting her into a monogynous colony, and vice versa. This turned out to be a complicated business requiring a modicum of patience, because obviously the workers took a dim view of it and starting roughing up the intruder. As soon as the first signs of aggression were apparent, we put the whole colony into the fridge so as to slow their rhythm of life and thereby cool down the fighting spirit of the attackers. After the nests were moved back and forth a few times between the cold of the fridge and the warmth of the lab, they settled down and accepted the foreign queen. As a result, the proportion of BB and Bb workers started to change with the replacement of resident workers by offspring of the new queen.

These experiments enabled us to demonstrate that there is a threshold effect and that a greater or lesser proportion of heterozygotic workers can turn the whole social organization of a colony upside down. If a nest contains more than 5 to 10 per cent Bb workers, it will be polygynous; under 5 to 10 per cent, it is monogynous.

The most visible and spectacular effect of this change in behaviour can be seen in BB workers in monogynous colonies. Left to themselves, they never accept the presence of any queen but their own. But as soon as a colony's population of heterozygotic workers exceeds 5 to 10 per cent, the original inhabitants will go through a complete change of attitude. They become hospitable and have not the slightest objection to the arrival in their nest of several Bb queens.

The same goes for polygynous nests. At the outset of our experiment, the nest contained about 60 per cent of heterozygotic workers (Bb), all the rest being BB. We then started to change things by inserting a queen from a monogynous society, the consequence of which was that the percentage of heterozygotic workers gradually decreased, dropping eventually to 5 to 10 per cent after three to four months. As soon as the crucial threshold was reached, the colony changed its mode of functioning, became monogynous instead of polygynous and so stopped admitting any new queens.

This shows that an ant's behaviour is dictated not only by its own genotype but also by the genotype of its nestmates. The organization of colonies is determined by interaction between genetics and social environment. That is to say, the b allele of the Gp-9 gene induces 'selfish' behaviour in the ants that carry it; but it does not prevent them from being responsive also to the others among whom they live.

28
The genomics of behaviour

The genetic inheritance of the red imported fire ant probably contains few genes that have as great an influence on the social organization of colonies as Gp-9. But it cannot be doubted that there are many others which also play a part, albeit less marked, in their behaviour.

It is those genes which we now have to seek out, benefiting from advances in the new science called genomics. The genome of *Solenopsis invicta*, unlike that of many other living organisms, such as bacteria, yeast, mice, even human beings, has not yet been completely deciphered. In the meantime it is already possible to find out the expression of numerous genes through the use of 'DNA chips'.

So, what are DNA chips? At this point it would probably be useful to pause and consider some basic facts about DNA. Genes, aligned on chromosomes, are in fact portions of DNA (deoxyribonucleic acid). Each gene is as it were a programme for making a particular protein, sometimes several proteins. The information is stored in it as an alphabet of four letters: A for adenine, C for cytosine, T for thymine, and G for guanine. It is the sequence of these four 'bases', as they are known in biology, which determines the composition of the protein. The transcription of the genetic code into its final product is not direct, as it must pass through a series of stages. The DNA, which is located in the nucleus of the cell, is first copied into another molecule, known as messenger RNA (ribonucleic acid). The RNA then leaves the nucleus and migrates to another part of the cell, the ribosome, where it 'translates' the words written with the four letters ACTG into amino acids, which are the building blocks of proteins. The proteins themselves are eventually synthesized through the action of another RNA, known as 'transfer RNA'.

Thus the genetic code can be compared to a cookery book with a recipe on each of its pages. It is rather like a set of very precise instructions which, if followed to the letter, produce a concrete result. This result, unlike the ones brought about by following real recipes, is neither a veal blanquette nor a cheese soufflé, but the setting in motion of the cellular machinery to make the different proteins which are the molecules essential to all living organisms.

Organisms contain, within the nucleus of each of their cells, a complete copy of their genetic inheritance. It is rather as though all cooks were in possession of the same book containing a complete collection of recipes. But just as a pastry-cook never has to look up the chapter on braising meat, no muscle needs digestive enzymes to work properly. So genes that are not required for an organ, a physiological function, or a particular behaviour lie dormant: they are not 'expressed'.

From genetics to genomics

About twenty years ago, scientists began using the tools of molecular biology to decipher genome sequences in many living species. To begin with, they were like archaeologists reading hieroglyphs before the Egyptologist Jean François (1790–1832) Champollion came along: they could decipher the words all right, but they couldn't make sense of them. So they decided to progress from reading to semantics, with the aim of ascertaining the conditions in which a gene is expressed, which is a good way of understanding what it is for. This task proved to be arduous and for a long time each team of scientists devoted their energy to analysing a mere handful of genes. These days things are different; it has become possible to study simultaneously, and compare, the expression of several thousand genes. Science has gone from genetics to genomics.

This change of scale can be attributed to DNA chips, a wonderful invention dreamed up at Stanford University (California) in the 1990s, which saves vast amounts of research time. Except for their size, there is no similarity between DNA chips and silicon chips. DNA chips are made of a fine glass slide, specially treated, to which are affixed the genes of the organism under study, or, more precisely, one strand of DNA, since the double helix of the DNA has already been unzipped. The operation being automated, it is possible to set out several thousand different genes in a predetermined order on a surface of several square centimetres. Their eventual role is to serve as probes or as hooks.

In order to ascertain which genes are active in an organism at any given moment, scientists take samples of messenger RNAs from the specimen being studied (their presence reveals the activity of the gene that they translate). The RNAs are transcribed as strands of DNA, marked as such with fluorescent substances, then put on the slide. Some strands of DNA from the sample being tested will hybridize with the genes on the slide. They do not do this at random: the hooks only catch strands that are complementary to themselves, in other words ones with which they can close up the DNA zipper and which thus originate from the same gene as they do. All that remains to be done is to read the finding with a scanner that analyses the fluorescence of the different genes on the glass slide—and Bob's your uncle. Red spots indicate, for example, activated genes and green ones genes that are inhibited.

This method is quick and effective. It is convenient in another way too: one does not have to decipher the genome of an organism in its entirety to ascertain the expression of some of its genes. So DNA chips are a great tool for studying the genetic origins of the behaviour of the fire ant *Solenopsis invicta*. We have already identified about 10,000 of the 12,000 to 14,000 genes and can study their expression.

This will open up new areas for research. Genomics might well be of assistance, for instance, in identifying the genes that influence the phenotypes of ants, their behaviours, or their smells. Genomics should also help enlighten us about the mysterious ways in which caste is determined. For the fact is that all larvae are born with similar genes, yet we still do not know with any precision the genetics of the process that makes some of them turn into queens and the rest into workers. Nor is it known why some of the latter specialize in domestic chores while the role of others is to defend the territory. What is apparent is that some genes, by being expressed a lot, a little, or not at all, are enough to influence what becomes of baby ants. Which genes do this remains to be seen. Given, too, that queens live much longer than workers, a genetic comparison between them might lead to the discovery of genes for longevity.

Suffice it to say that with DNA chips we can now investigate genes which, like Gp-9, have a function in the social organization of insects or an influence on their way of life. What is now on the cards is the development of the genomics of behaviour.

Inborn or acquired?

Words like 'behaviour genes' always raise a few hackles. They smack of the terminology used by some scientists who have jumped to conclusions about the so-called 'gene for intelligence' or 'gene for violence' in humans. Such an unseemly rush to judgement obscures the fact that characteristics like intelligence and violence are too complex to be reduced to the expression of a single gene. It also ignores the role of people's backgrounds—their upbringing, the socio-economic setting in which they live, the things and people they encounter, the whole process of acquired experience which, by interacting with what is inborn in us, turns us into who we are.

Some of the hostility to the notion of the genetics of behaviour derives from quite understandable fears; and there can be no doubt that, if a particular human behaviour could be shown to have a genetic component, this might have dangerous implications. Some might be persuaded by the argument that because this or that behaviour is genetic, it can't be helped, that it is pointless to provide any social or other assistance to someone who is, for example, genetically predestined to aggression. There are no grounds for accepting such ideas, given that character traits, even if influenced by heredity, are invariably also affected by external factors, and whatever is acquired can always be modified. So treating people as no longer equal cannot under any circumstances be justified by genetics.

All that said, facts are facts. It cannot be denied that most behaviours, whether individual or common to a species, do have a genetic component. At times a single gene can suffice to change a way of life, as can be clearly seen in Gp-9's ability to condition unaided the monogynous or polygynous organization of colonies of *Solenopsis invicta*. Nor are these ants the only creatures in which this sort of thing can occur: science has recently discovered that changes to a gene can be enough to make male voles either monogamous or polygamous, demonstrated by the fact that by tampering with the particular gene, we can produce a variation in the number of the males' mating partners.

Mostly, however, a behaviour is governed by numerous genes interacting with each other. At present, the influence of genetics on behaviour has been shown to occur in some living organisms; and it can be quite confidently predicted that there will soon be many more such discoveries. And some of these will undoubtedly be made by myrmecologists, for ants happen to be excellent subjects for this type of research.

29

So what's so special about the genome of fire ants?

As we have already said, there is no need to know the whole story of ants' genome to investigate the genetic basis of their behaviour, though the task would certainly be greatly facilitated and accelerated if we did know the complete repertory of their genes, in other words if their genome was deciphered.

Deciphering it would not in fact be impossible. Quite a few genomes, some of which belong to living organisms that are much more complex than an ant, for instance chickens, chimpanzees, and even humans, have already been mapped. Compared to the three billion 'letters' making up human genome, the 800 million or so likely to be found in the DNA of ants look like small beer indeed. Reading them, however, will require the bringing together of many different skills and a substantial amount of funds.

At present, the genomes of only a small number of highly social species, human beings, chimpanzees, and bees, have been deciphered, so it would be nice to be able to add ants to the list. Because they are the most social of social insects, they are the perfect model for studying links between genetics and behaviour.

Furthermore, a comparison of their genome with that of bees should make it possible to identify certain genetic features underlying the development of sociality in insects to begin with, then in vertebrates.

Close investigation of the DNA of an ant is thus required. But which ant? The most logical choice would be the fire ant *Solenopsis invicta*, the worst type of invasive species. Traditional methods for halting their advance having failed, genomics could possibly lead to the discovery of other ways of dealing with them. This might help us to identify the genes that govern reproduction or perhaps even those that make it possible for social harmony to exist inside colonies, thus giving us a weapon against them. If it should prove possible, say, to prevent the queens from reproducing or to affect the mechanism of caste determination, this might well offer some hope of exterminating the reproductive caste, thereby destroying their colonies.

There are further reasons for taking an interest in the genome of *Solenopsis invicta*. Because of the damage they cause, these ants have already been thoroughly studied. All over the world, dozens of teams of myrmecologists have investigated their biology, habits, and behaviours, making it one of the ants best known to science. So it would appear logical to take things further by analysing the totality of its genome. It should be added that the main function of the red imported fire ant would be as a model, since the deciphering of their genome is expected to supply information applicable to many other species.

Genes and castes

Another benefit of sequencing the genome of *Solenopsis invicta* would be that it could help us to understand the origin of polymorphism in ant colonies. It was recognized long ago that queens are unlike their workers: they are larger and they have sexual organs. It is also known that there are great morphological differences among workers in some species, including as it happens *Solenopsis invicta*. The occurrence of several morphological castes is the source of much speculation. What is the process that turns a larva into a queen rather than a worker? What are the mechanisms that make one worker a normal worker and another a soldier? We know that the way the workers feed the brood has an influence on the development of the larvae and that, though this holds good for most species, there are some exceptions to the rule. In *Pogonomyrmex*, for example, it is now clear that genetic factors and maternal effects, including changes made by females to the hormonal content of their eggs, also influence the process of caste determination. Identification of which factors they are could enlighten us on how genes and environment interact during the development of individuals.

Great things are also expected from a comparison of the fire ant genome not only with that of another social insect, the bee, but with that of the parasitic wasp *Nasonia*, which is solitary. Analysis of genetic features shared or not shared by these three Hymenoptera would provide a good way of identifying potential genes for sociality, some of which it is not fanciful to think may also be present in human beings.

Understanding ageing

It might be thought that the gulf between ants and humans is so vast as to be unbridgeable. But, genetically speaking, the specificity of our own species is really quite slight: we share 99 per cent of our genes with chimpanzees and about 40 per cent with mice! The future will tell what similarities there are between our DNA and the DNA of fire ants. One thing is already established, that there are features common to both humans and ants; and identification of them could improve our understanding of some biological mechanisms, first and foremost those involved in human ageing.

Understanding why and how living things age has become a real challenge for scientists in many different disciplines. Specialists in evolution would like to know what underlies the enormous disparities in the longevity of living creatures, why, for instance, some turtles can live for 100 years, while the life expectancy of mice is at best three. Biologists and doctors, too, take an interest in this: faced with the conundrum of ageing in industrialized societies, they would be glad to be able to delay or even counter the effects on organisms of living longer. They are investigating the cellular, molecular, and genetic origins of senescence, work in which they focus on model organisms. Hitherto, they have relied essentially on the worm *Caenorhabditis elegans*, the fruit fly *Drosophila melanogaster*, or the yeast *Saccharomyces cerevisiae*, all of which are well known and easy to study. But all of these organisms have a very brief lifespan, which is why they are not ideal for studying the mechanics of ageing.

Ants would be much better suited to this purpose. For one thing, their longevity can be remarkable (some queens have been known to reach the age of twenty-nine); and on average they live 100 times longer than most other insects. For another thing, within a single species there are wide variations among the different castes: a queen, for instance, may live 500 times longer than the males and twenty times longer than her workers. Disparities of this sort can even be apparent within the worker caste, something that we have demonstrated with Michel Chapuisat in weaver ants, among whom lifespan is a function of specialization, little nursemaids living statistically a few weeks longer than the large hunter-warriors. What is especially surprising about this finding is that, in most animal species, smaller individuals do not live as long as larger ones. Among weaver ants, therefore, the tendency is inverted, even when colonies are housed in a laboratory, away from natural dangers. This must mean that differences in gene expression between weaver ants of different size directly affect the difference in their longevity.

The fact remains that queens and their worker daughters share a good part of their genome, as do the workers of any given colony. The differences observed in their longevity cannot therefore be explained by the nature of their genes, but rather by the way the genes are expressed, that is 'activated' by external and environmental factors. This hypothesis could be proved, or perhaps disproved, by analysis of the fire ant's genome, added to which is the possibility that, in a more general way, deciphering the genome could also reveal the existence of genes linked to ageing that might be peculiar to organisms with long lifespans.

Walking chemical plants

The implications could go even further, in that there could be medical benefits to be derived from the genome of the fire ant, first and foremost the development of new antibiotics. After all, fire ants live in an environment favourable to the growth of pathogens, the humid warmth of the nest providing an excellent culture medium. In addition, their living conditions inside the nest entail much close bodily contact and they often exchange food with each other. This way of life is ideal for transmitting microbes throughout the colony—despite which, and despite being susceptible to certain diseases, fire ants, like all other ants, seem to have developed protection against micro-organisms, probably because of the substances secreted by their glands and their stings. So it is entirely possible that work on these walking chemical plants could produce recipes for new bactericides or even fungicides, herbicides, and insecticides.

Another source of inspiration might be the venom used by *Solenopsis invicta* for a variety of purposes. They use it, for example, to defend their nest against predators, vertebrate or invertebrate, and for keeping other ants out of their territory. It also serves as a toxin for overcoming their prey and even, as it appears, as an antiseptic against the growth of micro-organisms in the soil which might infect the nest. Not bad going for a single substance! It is also a further reason, if further reasons are required, for trying to understand the mechanisms via which the ants can synthesize this do-all venom. Here, too, information from their genome would be likely to lead to a breakthrough.

It can be seen that the full sequencing of the *Solenopsis invicta* genome would provide enough data for scientists in many different disciplines to work on for many years. Entomology for one, of course, would make great strides. Once deciphered, the genome of fire ants would provide new tools to elucidate some of the mysteries surrounding their lives and their exceptional social organization.

Part VIII

High-tech Ants

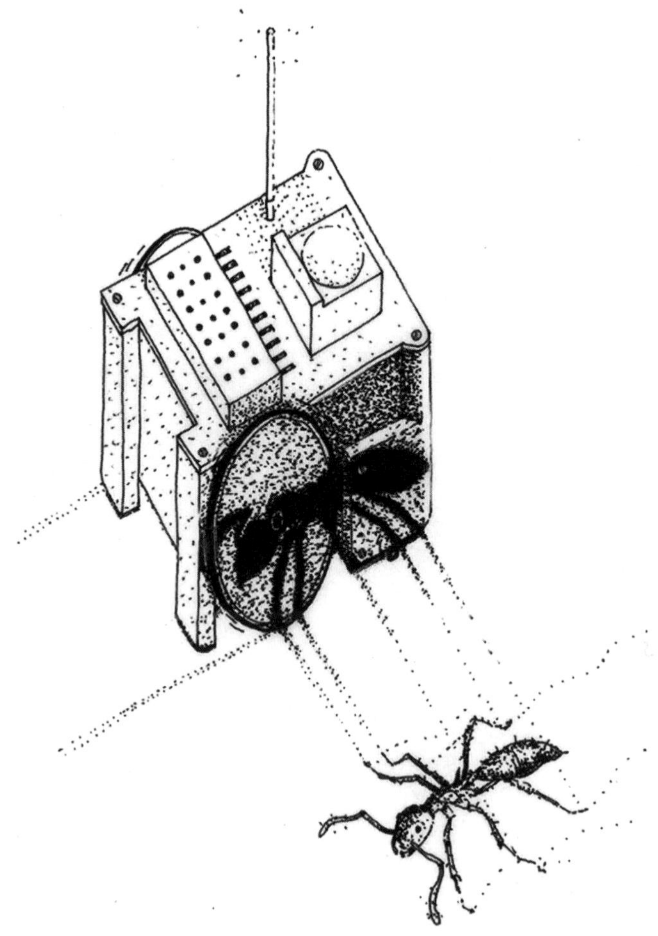

Drawing 8 Ant robots Engineers are devising teams of mini-robots based on the social behaviour of ants.

30
Computer-modelling behaviour

That a great many entomologists take a close interest in ants, in their behaviour and their genes, is only to be expected. But it may be more surprising that engineers such as computer scientists and specialists in robotics should also find them of interest and even try to borrow ideas from their social arrangements. The fact is that experts in artificial systems are now drawing upon our knowledge of ants to help them in their development of automated devices for speeding up telecommunications or even for designing teams of robots to send to Mars.

At first sight, there is nothing in common between the future miracles of the technologists and creatures that appeared on earth millions of years ago. Nor are ants, with their tiny brains, gifted with great intelligence. Workers, taken individually, spend their time doing very simple and repetitive tasks and can even act in completely disorganized ways, not infrequently undoing the work done just a few moments before by a nestmate. Nevertheless, collectively, they are capable of great achievements. They can construct nests of sophisticated architecture which are veritable cathedrals on an entomological scale; and they can come up with solutions to complex problems.

Yet, in accomplishing these tasks, they are following no pre-established design. Each ant acts with complete autonomy, for theirs is a society without hierarchy, the queen, despite the title we give her, having nothing to do with the allotment of jobs. So cooperation within the colony is self-organizing. 'In insect societies,' says Jean-Louis Deneubourg from ULB, 'the overall "plan" is not explicitly programmed in individuals, but rather is an outcome of the linking together of a great many elementary interactions between individuals or between individuals and the environment. What there is, in fact, is a collective intelligence constructed out of a multitude of separate and simple minds.' This phenomenon is known as 'swarm intelligence'.

Model cemeteries

Thus collective and consistent behaviour arises out of multiple individual acts that are apparently disorganized. The intriguing question that arises is how can such order derive from such chaos? An international team, under the direction of Guy Théraulaz from the Research Centre on Animal Cognition at the Université Paul-Sabatier (Toulouse), supplied a partial answer by putting the behaviour of the harvester ant *Messor sancta* in an equation and computer-modelling it.

As a measure of public hygiene and so as to reduce the risk of infection in the colony, workers of many species are careful to remove dead bodies from the nest. Alerted by the chemicals given off by corpses, notably oleic acid, the 'gravediggers' pick up the dead insects and deposit them outside the nest. But they do not just push them out at random; they take care to leave them in neatly arranged heaps, thereby making veritable 'cemeteries'.

With the aim of making a close analysis of the activities of these undertakers, the team made an arena in their laboratory and littered the outer rim of it with dead ants. When they let workers of *Messor sancta* into this space, they saw them assemble little clusters of bodies in the space of a few hours, some of which they then disassembled so as to add the bodies to some of the other clusters. By the end of the whole process, all that was left was a small number of heaps regularly spaced out. As Théraulaz and his colleagues put it, 'a large-scale regular spatial pattern eventually emerges from the individual carrying activity of each ant'.

Observation was followed by computer-modelling. By studying the video of the arena experiment, the team were able to determine the parameters governing the individual behaviours of each ant. In particular, they calculated the probability for a worker to either remove a corpse from one of the piles or add one to it, going on the size of whichever pile it was dealing with. From these data they derived statistical laws showing the average behaviour of a great number of ants in identical circumstances.

This study showed that the probability of an ant adding a corpse to a pile has nothing to do with the length of time it has been carrying it. Thus fatigue is not the reason it does so. It also became apparent that any isolated corpses were quickly picked up. It was observed that the bigger a cluster of corpses was, the less likely the ants were to take any from it and the more likely to add others to it. This does not mean that an ant has the ability to estimate the sizes of the different clusters, but simply that the larger the heap, the more time it spends in negotiating it and the greater the chance that it will drop its burden there. Thus the taking of corpses from a cluster stimulates the taking of more corpses, just as the adding of them stimulates the adding of more; and when any small change in the size of a cluster becomes apparent, it soon becomes cumulative, leading to the disappearance of one of the other clusters.

These observations and painstaking calculations eventually resulted in the definition of the two mechanisms responsible for the growth of the heaps of bodies. One locally activates the adding of bodies and the other, farther-reaching, inhibits the taking of them, bodies already added not being available for the formation of new piles. In this amazing collusion between entomology and mathematics, ants provide unexpected support for Turing's model. In the 1950s, Alan Turing, the British mathematician seen as the father of modern computing, propounded laws enabling prediction of the appearance of spatial structures during certain chemical reactions. When two coloured substances acting on one another are placed in a milieu where they diffuse, patterns such as streaks or hexagonal shapes appear spontaneously. One of the substances is known as 'activating' and the other as 'inhibiting', which is how chemistry leads us straight back to the ants. Activators favour their own production, whereas the others inhibit the production of the activators. The mechanisms posited by the Turing model (short-term activation and long-term inhibition) are thus the same as those shown by ethology as explaining the behaviour of the *Messor* undertakers.

Turing's model had been drawn on before by biologists to explain the formation of patterns on fish and shellfish or the stripes and other marks on the coats of zebras and leopards. This interpretation had, however, been disputed, as it was unsupported by any conclusive experimental demonstration and no underlying mechanism had been shown. Thanks to the French scientists and their colleagues, not to mention the ants, this support has now been provided. For the first time, the laws posited by Turing have been validated in the world of living things. Here is a sign that entomology and mathematics make good bedfellows.

31
Of ants and IT men

Alan Turing probably never imagined that ants might one day turn out to be a source of confirmation of his ideas, but his descendants, the computer scientists of our own day, have certainly grasped what a valuable resource ants are. Bio-inspiration being the order of the day, some computer specialists have deliberately chosen to use ants in their models.

This turns out to be a genuinely inspired choice, for it has helped them solve a real mathematical conundrum: the 'travelling salesman problem', which consists of finding the shortest route for visiting no more than once each of a set of towns linked by roads. It looks quite easy; but in fact it is far from simple, for even if the traveller has to visit only fifteen towns, he has to choose between ninety billion possible routes. Now, haven't we come across something like 'finding the shortest route' before? Remember the great experiments carried out by Jean-Louis Deneubourg and colleagues of ULB which made sense of Argentine ants' liking for shortcuts (see p. 63). Their finding was that the shorter a path was between the nest and a food supply, the more trips out and back the workers could make, thus increasing the amount of pheromones deposited on it, which made the foragers prefer that path.

In an attempt to solve the travelling salesman problem, Marco Dorigo and his team, also from ULB, did a computer simulation of the strategy used by real ants. They used virtual insects which left digital pheromones on their paths. All they had to do was start up the programme and send out their virtual scouts along the electronic circuits. They made sure that their ants 'knew' in advance the distances between towns, so that they would favour the shortest itineraries.

The ants set out in random directions, then they turn back; and the shorter the distance they cover, the more pheromones they leave. Once all of them have returned to their starting point, it is easy to detect the paths that are richest in pheromones; and it follows that a greater number of shorter paths will figure among these. In the second stage of the experiment, the scientists once again released their dummy workers which, like their real counterparts, all set off along the paths already marked by pheromones, still favouring the links between towns nearest to each other. Having repeated this sequence a certain number of times so as to reiterate their calculations, all they had to do to find a short path was link up the most frequented ones. This short path was not necessarily the shortest one, but it was at least an almost optimal solution to the problem. In addition, as stated by Guy Théraulaz and Eric Bonabeau, who was at the time the director of the Eurobios company in Paris, the system is flexible: 'Since the artificial ants continually explore different paths, the various pheromone-laden ones offer back-up possibilities. Thus, if a path is blocked, alternatives are already in place.'

Programmes of this kind, directly inspired by the behaviour of ants, already have concrete applications that may seem improbable. For instance, the chemical company Unilever adopted a similar approach to organize movements between the storage tanks, mixers, and packing lines in one of its factories—swarm intelligence strikes again.

From telephones to the Net

Swarm intelligence might soon find its way into telephone systems. The effective management of communication systems is, after all, not child's play: if there are congested nodes in a network, they must be avoided and calls must be redirected to lines that are not so busy. In other words, the tracking of messages must be optimized.

If there is a problem with overloading, ants come to the rescue. The system developed by research engineers in laboratories at Hewlett Packard and the University of Bristol also uses digitized insects and their equally digitized pheromones. The virtual ants are programmed to leave their traces at certain nodes so as to reinforce the routing of messages through less busy areas. In parallel with this, an 'evaporation' mechanism makes it possible to thin out the traffic on the overloaded lines. Circulating through the network are ant-like digital agents which are instructed to react to the density of calls: if they go quickly from one node to another, it is a sign that the line is clear and they can deposit a sizeable amount of pheromone; if not, they deposit much less. In this way it is possible to continuously adjust the 'routing tables' at each node of the system which direct calls to their recipients. The ingenuity of this method was not lost on France Télécom or British Telecom, who were among the first organizations to take a close look at such bio-inspired software.

Obviously, the Internet also counts as a communication system. The circulation of data via the Web being even more unpredictable than telephone conversations, it is no mean feat to maximize flow and minimize delays. This is why Marco Dorigo and his team once again trotted out their virtual ants. They released them periodically at the network nodes, then, without telling them what their position was, asked them to go to a particular spot. A quick check of the pheromone map was enough to define the route that in each case linked one point to another most rapidly. Early results from these simulations suggest that this mode of traffic management is more effective than the protocols in use at present. Dorigo says he is convinced that this system will make for easier and faster surfing among the huge quantities of information available on the Net.

Identifying defaulters

Ants not only have a built-in liking for shortcuts, they also show a decided penchant for tidying, for putting like things together with like. This behaviour has been noticed by computer people who work for banks or insurance companies.

Just as the painstaking *Messor sancta* cemetery workers make clusters of corpses, so *Temnothorax unifasciatus* nursemaids have their own special way of sorting their broods: they set out the insects in concentric circles, depending on their stage of development. In the centre they put the eggs and smaller larvae; outside these they set the pupae, and all round the outer edge the larger larvae. That is, they classify the baby ants by size.

What interests bankers is of course not the size of their customers, but their ability to repay a loan. The question arises of how to easily identify potential defaulters. One good solution would be to proceed by analogy, grouping together people with similar characteristics, age, sex, marital status, type of housing, profession, favourite banking services, and so on. This would be a good way of identifying groups marked by a preponderance of defaulters, so that they could be required to meet more stringent conditions. Needless to say, banks and insurance companies already use this way of analysing clusters of customer profiles. However, they lack the software to enable easy visualization of the data obtained, which is where *Temnothorax unifasciatus* comes in. Let us imagine customers as eggs, pupae, or larvae large and small. Virtual ants, taking account of their profiles, could move them about and group them according to their similarities with the other borrowers. In this model, each debtor is represented by a dot; and the more any pair resembles another, the shorter the distance between the dots.

The ant-based approach not only makes the results of any calculation immediately obvious but, according to Théraulaz and Bonabeau, it also boasts 'an intriguing feature': 'The number of clusters emerges automatically from the data, whereas conventional methods usually assume a predefined number of groups into which the data are then fit. Thus, ant-like sorting has been effective in discovering interesting commonalities that might otherwise remain hidden.' Defaulters beware—the ants are on to you!

32
Swarm robotics

The term 'swarm intelligence', so apt a description of ant behaviour, was coined in 1989 by Gerardo Beni, a professor of electrical engineering at the University of California (Riverside). His original use of it, in the context of groups of bio-inspired robots, was itself inspired, for if there is an area where what is known as 'distributed intelligence' looks like having a great future, it is robotics.

This is a field in which the current trend is not only towards developing fancy machines bristling with detectors and capable of doing all sorts of things, engineers are also concentrating on the development of swarms of small robots with the ability to cooperate in accomplishing tasks that any one of them would be incapable of doing by itself. And rather than putting them under the command of a central computer, it now seems much smarter to let the group self-organize after the manner of social insects.

The idea was simultaneously launched in the early 1990s by the entomologist Jean-Louis Deneubourg and Professor Rodney Brooks of the Massachusetts Institute of Technology, the pioneer of autonomous robotics. According to Brooks, who is clearly a man who has done his myrmecological homework, the model of ant colonies and beehives shows their inhabitants to be so many robots. Individually, the members of these societies know little about the full task to be accomplished, such as the building of a hive. But once all the individuals work together, the task emerges from their interactions.

This was the beginning of collective robotics, with its double principle of a) leaving it up to each robot to act in accordance with simple behavioural rules, responding to its own appraisal of its environment, and b) programming communication and mutual assistance among the machines. There is of course no imitation of real ants with their chemical recognition signals. Instead, the colony of robots exchange their information via infrared mechanical signals or with colours and sounds.

For the last fifteen years or so scientists and engineers have vied with each other in imaginative and creative ideas for the making of mini-robot groups and giving them all sorts of tasks to perform. So far, none of these armadas has actually left the harbour of the laboratory; nor do concrete applications appear to be quite within reach. Nevertheless, there can now be no doubt that swarm robotics is technically feasible.

Nor is there any shortage of achievements. When collective robotics was still in its infancy, about fifteen years ago, a team of German, British, and Belgian scientists set about imitating *Messor* ants and their corpse-clustering, with robots able to collect disks scattered on the ground. At the time, the system was pretty rudimentary, in that the machines did not communicate with each other and could only go on the size of a pile of disks: the larger the pile, the easier it was for them to detect it and add a new item to it. This initial experiment was repeated a few years later by a Canadian team from the University of Alberta to show that several robots could work together, though still without communicating, to move a box.

In similar vein, Alcherio Martinoli and Francesco Mondada of the École Polytechnique Fédérale de Lausanne (EPFL) used a group of 'Kheperas' (miniature mobile robots just over five centimetres in diameter) to build a brick wall and pull a stick out of a hole in the ground.

Working with Michael Krieger and our colleagues from EPFL, we gave our Kheperas the task of 'leaving the nest', going to look for 'food', and bringing it back home. We added a further condition: the robots had to keep the overall energy of the nest above a certain level, which means they had to collect the maximum of 'food' while expending the minimum of effort. These little machines on wheels, communicating via infrared signals, functioned like their living counterparts. Each of them went about its business, moving randomly until it found a 'seed', which it would then pick up with its arm and take back to its starting point. If a robot came upon a plentiful source of food, it memorized the way back to the nest and as soon as it got there it recruited nestmates and led them to the feeding ground. It was observed that productivity increased proportionately with an increase in the number of worker robots, but only up to a point, beyond which there were risks of collision and disorientation. This happens to be exactly what takes place in a colony of ants when the numbers of workers trying to do the same job becomes unmanageable and there has to be a change of strategy.

In the experiment, the mathematical algorithms, which function in a way as the 'brain' of the machine, predefined the actions of the robot. To let the behaviour of their robots evolve, Francesco Mondada and Dario Floreano, the Director of the Intelligent Systems Laboratory at EPFL, improved them by endowing them with neuronal systems and genetic algorithms. Recently, working with Dario Floreano and other colleagues, we took a step further, in an attempt to understand the mechanisms of cooperation which arise within an artificial society of robots. To that end, we made an artificial selection of the most efficient individuals, crossing their 'genomes' exactly as breeders do in their quest for higher-performance animals.

To begin with we selected by digital simulations via computer, which led to the rapid production of many generations of artificial creatures. Going on the results thus obtained, the engineers at EPFL constructed about ten real robots which have already demonstrated an ability to go about things in a coordinated way and help each other to accomplish the tasks we set them. These preliminary studies also show that, in order for the machines to cooperate and display altruistic behaviour, they must be 'relatives' of each other, that is to say they must share quite a lot of their algorithms—exactly like their myrmecological models!

Weaver ants on wheels

Mondada and Floreano have continued pursuing related lines of research. Clearly fascinated by ants, they have been especially attracted to weaver ants and tempted to imitate their ways of holding on to each other as they build nests in the canopy of tropical forests. So they built little robots with arms which hold on to one another. These 'swarmbots' (the word is a conflation of 'swarm' and 'robot') have the ability to form themselves into chains, circles, or squares and can help each other to jointly overcome obstacles. In one of their groupings, by pushing and pulling one another, they manage to climb stairs.

The scientists at EPFL have even designed an amusing experiment to demonstrate that if a few of their mini-machines join forces (they are about ten centimetres in diameter and weigh 700 grams), they can actually move loads thirty times heavier than they are. Mondada likes showing visitors to his laboratory a video of a little girl lying on the ground and a cohort of well trained swarmbots. They link up in four chains of five robots each and take up position on the child's left side. The leading robot in each chain then takes hold of her clothing, at which she looks rather worried, and all of them pull gently, managing to shift their load by a few metres. This demonstration is pure spectacle; but it also makes the point that 'the concept is feasible', as Mondada puts it.

It is also a way of proving that communication can happen between robots. Not only are they equipped with cine-cameras, but their 'bodies' are also surrounded by a ring that changes colour in accordance with whatever information they wish to transmit to their fellows. For instance, red means, 'I need help'; and any robots that come to the rescue start beeping as though saying, 'Don't panic, I'm coming.' As Mondada explains, 'Robots could communicate through messages sent via the Net using radio signals. But that wouldn't give them information about their respective positioning.' It is better to use colour and sound, the veritable pheromones of ant-robots.

Altruists or warriors

Autonomous mini-robots able to adapt to changes in their surroundings, capable of completing a mission even if one of them breaks down, offer ideal systems for foraging in hostile environments or for acting in situations where an ability to respond to the unexpected is required. One can easily imagine squads of ant-robots exploring the surface of the moon or Mars. They could climb over large stones, they could bridge crevasses, all with much less difficulty than 'Spirit' and 'Opportunity', the NASA robots which had some trouble moving about the surface of the red planet. NASA has actually shown an interest in bio-inspired machines and has plans for sending them into space for the purpose of assembling large structures.

Might it not also be possible to hope that one day such robots could prove useful in cases of catastrophe? If they were deployed in places devastated, say, by an earthquake or a tsunami, they could take part in the humanitarian aid effort. Because they are so tiny, albeit not as tiny as ants, they could easily be sent into the debris where they could link up with each other to climb over obstacles or holes, and to assist in finding survivors.

Obviously, machines like these would also have great appeal for the military. Engineers at iRobot, the firm set up by Rodney Brooks, are already at work on developing a team of small automata capable of patrolling inside buildings and defusing any explosive devices they might find. American scientists from the Defense Advanced Research Projects Agency (DARPA) are taking a close interest. In 1998, they challenged roboticians to come up with tiny user-friendly reconnaissance robots that could be part of troops' basic kit. To make things interesting for them, DARPA devised a dramatic scenario: a group of terrorists seizes a building and takes all the people inside hostage (they have of course blocked all entrances and masked all the windows); but what they don't know is that an army of little robots, all equipped with microphones and cine-cameras, or even chemical or biological detection devices, has been surreptitiously introduced via the ventilation system and is busy informing the security forces in real time about what the hostage-takers are up to. Based on the information, an attack is launched and the siege is brought to a satisfactory conclusion with the capturing of the terrorists before they can even prime their weapons.

This scenario is an imaginary one, closer to science fiction than to the present state of progress in robotics technology. But there is no shortage of such scenarios, for ideas tend to outrun reality when potential applications of swarm robotics are being thought about, including the most unexpected. These include turning swarm robots into sheepdogs with the ability to control a flock of ewes, or into chickens able to prevent outbreaks of collective panic in poultry farms. However, before we reach that stage, there are quite a few technical difficulties to be solved.

In the meantime, it is quite possible that swarms of bio-inspired robots could be used in biology as aids to understanding how animals adapt to new environments. This is no easy task, for such studies with living organisms are difficult. So far, the only ones found suitable are bacteria, which reproduce rapidly and are easy to cultivate in the laboratory. Nevertheless, such microorganisms do not make ideal models for this type of work. Ants would be much better, though it is much trickier to breed them and study the way their behaviour evolves over many generations. Robots, on the other hand, present many advantages: being artificial, they can not only be made to adopt the behaviours of social insects, they can be manipulated, selected, or mutated ad infinitum. They might even turn the tables on myrmecologists and end up as models for their models. First there were bio-inspired engineers; soon we may come full circle with technologically stimulated entomologists.

Conclusion

Since time immemorial, human beings and ants have lived side by side, usually getting on well together. Our distant ancestors, living as they did in close contact with the natural world, could not help being impressed by the tiny creatures.

It was not until the eighteenth century, however, with the beginnings of the study of natural history, that the observation of ants can be said to have started developing into a science. The earliest practitioners of this craft closely observed the lives of ants and gave accurate descriptions of many of their ways. They were the first to genuinely understand what was so remarkable about ant societies; they described the caste structure, the organization of work, the modes of communication; and they drew attention to the altruism of the workers. The trail they blazed has been followed by a multitude of entomologists who have gone on to analyse in ever increasing detail the magical and many-faceted world of ants.

Our own time is one not only of observation and description but also of experimentation and explanation. Because we now know how ants reproduce, we are beginning to understand why workers behave as they do; because we have looked right into their genes, we are able not just to explain the wheeling and dealing, the power struggles, which go on inside colonies, but also to predict alliances and conflicts. And this is only the beginning, for myrmecology is an area in which investigative genetics provides a tool that has not yet been fully utilized.

Ants are of course immensely interesting insects. But over and above the interest that life in their universe can inspire, they offer another attraction, in that they can serve as models for many areas in the life sciences. They can be of assistance, for example, in the study of various animal societies and in helping us to understand the evolutionary processes through which our own species became a social one. They can contribute to the deciphering of ageing mechanisms or the development of a reliable genomics of behaviour in living creatures, including human beings.

The long-standing connection between ants and humans is nowhere near ending. No one knows quite what direction it will take, but one thing is certain: creatures as fascinating as ants still have plenty of surprises in store for us.

Further Reading

Classics

Forel, A., *Le monde social des fourmis* (Librairie Kundig, 1921; English trans., *The Social World of the Ants Compared with That of Man*; London and New York: G. P. Putnam, 1928).

Latreille, P.-A., *Essai sur l'histoire des fourmis de la France* [1798]; reissued (Paris and Genève: Champion-Slatkine, 1989).

Wheeler, W. M., *Ants: Their Structure, Development and Behavior* (New York: Columbia University Press, 1965).

Fiction

Calvino, I., 'The Argentine Ant', in *The Watcher and Other Stories*, trans. William Weaver (New York: Harcourt Brace, 1971).

Werber, B., *Les fourmis* (Paris: Albin Michel, 1991; English trans., *Empire of the Ants*, New York: Random House, 1999).

Websites

General information on ants

<http://www.antcolony.org/>

<http://antfarm.yuku.com> (forum)

<http://www.myrmecology.info/index2.html>

<http://alpha.zimage.com/~ant/>

Picture galleries

<http://www.myrmecos.net/>

<http://ant.edb.miyakyo-u.ac.jp/> (Australian and Japanese ants)

<http://www.tightloop.com/antsite/default.php> (ants of SW USA)

Taxonomy and distribution

<http://antbase.org/> (taxonomical datebase)

<http://www.antweb.org/> (ants of America & Madagascar)

<http://www.antbase.net/> (ants of Borneo, Malaysia & Germany)

<http://www.ento.csiro.au/science/ants/> (Australian ants)

Specialized

On ants in general

Bolton, B., 'Synopsis and classification of Formicidae', *Memoirs of the American Entomological Insititute*, 71 (2003): 1–370.

Bourke, A. F. G., and Franks, N. R., *Social Evolution in Ants* (Princeton, NJ: Princeton University Press, 1995).

Crozier, R. H., and Pamilo P., *Evolution of Social Insect Colonies. Sex Allocation and Kin-Selection* (Oxford: Oxford University Press, 1996).

Gordon, D., *Ants at Work: How an Insect Society is Organized* (New York, W. W. Norton, 1999; Free Press, Simon & Schuster, 2000).

Hölldobler, B., and Wilson, E. O., *The Ants* (Berlin: Springer-Verlag, 1990).

——*Journey to the Ants: a Story of Scientific Exploration* (Cambridge, MA: Harvard University Press, 1995).

Passera, L., and Aron, S., *Les fourmis: comportement, organisation sociale et évolution*, Publication du programme de monographies du CNCR (Conseil national de recherches du Canada), 2006.

On ecological success and environmental impact

Folgarait, P. J., 'Ant biodiversity and its relationship to ecosystem functioning: a review', *Biodiversity and Conservation*, 7 (1998): 1221–44.

MacMahon, J. A., Mull, J. F., and Crist T. O., 'Harvester ants (*Pogonomyrmex spp.*): their community and ecosystem influences', *Annual Review of Ecology and Systematics*, 31 (2000): 265–91.

Wilson, E. O., *Success and Dominance in Ecosystems: the Case of Social Insects* (Oldendorf: Ecology Institute, 1990).

History and primitive ants

Brady, S. G., Schultz, T. R. Fisher, B. L., and Ward, P. S., 'Evaluation of alternative hypotheses for the early evolution and diversification of ants', *PNAS*, 103 (2006): 18172–7.

Grimaldi, D., and Agosti, D., 'A formacine in New Jersey Cretaceous amber and early evolution of the ants', *PNAS*, 97 (2000): 13678–83.

Moreau, C. S., Bell, C. D., Vila, R. Archibald, S. B., and Pierce, N. E., 'Phylogeny of the ants: diversification in the age of angiosperms', *Science*, 312 (2006): 101–4.

Peeters, C., and Ito, F., 'Colony dispersal and the evolution of queen morphology in social Hymenoptera', *Annual Review of Entomology*, 46 (2001): 601–30.

Schultz, T. R., 'In search of ant ancestors', *PNAS*, 97 (2000): 14028–9.

On life in society and division of labour

Beshers, S. N., and Fewell, J. H., 'Models of division of labor in social insects', *Annual Review of Entomology*, 46 (2001): 413–40.

Chapuisat, M., and Keller, L., 'Division of labour influences the rate of ageing in weaver ant workers', *Proceedings of the Royal Society of London*, Series B: Biological Sciences, 269 (2002): 909–13.

Robinson, G. E., 'Regulation of division of labor in insect societies', *Annual Review of Entomology*, 37 (1992): 637–65.

On communication

Ayasse, M., Paxton, J. R., and Tengö, J., 'Mating behavior and chemical communication in the order Hymenoptera', *Annual Review of Entomology*, 46 (2001): 31–78.

Baroni-Urbani, Buser, C. M. W., and Schilliger, E., 'Substrate vibration during recruitment in ant social organization', *Insectes sociaux*, 35 (1988): 241–50.

Devigne, C., and Detrain. C., 'Collective exploration and area marking in the ant *Lasius niger*', *Insectes sociaux*, 49 (2002): 357–62.

Dussutour, A., Fourcassié, V., Helbing, D., and Deneubourg, J.-L., "Optimal traffic organization in ants under crowded conditions", *Nature* 428 (2004): 70–73.

Lewis, T., *Insect Communication* (New York: Academic Press, 1984).

Robinson, E. J. H., Jackson, D. E., Holcombe, M., and Ratnieks, F. L. W., ' "No entry" signal in ant foraging', *Nature* 438 (2005): 442.

On family models

Monnin, T., and Peeters, C., 'Dominance hierarchy and reproductive conflicts among subordinates in a monogynous queenless ant', *Behavioral Ecology*, 10 (1999): 323–32.

Ross, K. G., 'Molecular ecology of social behaviour: analyses of breeding systems and genetic structure', *Molecular Ecology*, 10 (2001): 265–84.

On parasites

Aron, S., Passera, L., and Keller, L., 'Evolution of social parasitism in ants: size of sexuals, sex ratio and mechanisms of caste determination', *Proceedings of the Royal Society of London*, Series B: Biological Sciences, 266 (1999): 173–7.

Buschinger, A., 'Evolution of social parasitism in ants', *Trends in Ecology and Evolution*, 1 (1986): 155–60.

Lenoir, A., D'Ettorre, P., Errard, C., and Hefetz, A., 'Chemical ecology and social parasitism in ants', *Annual Review of Entomology*, 46 (2001): 573–99.
Topoff, H., 'Slave-making queens', *Scientific American*, 281 (1999): 84–90.

On army ants

Franks, N. R., and Fletcher, C. R., 'Spatial patterns in army ant foraging and migration: *Eciton burchelli* on Barro Colorado Island, Panama', *Behavioral Ecology and Sociobiology,* 12 (1983): 261–70.
Gotwald, W. H. Jr., *Army Ants: the Biology of Social Predation* (New York: Cornell University Press, 1995).

On weaver ants

Hölldobler, B., and Wilson, E. O., 'Weaver ants', *Scientific American*, 237 (1977): 146–54.

On ant navigation

Chameron, S., Beugnon, G., Schatz, B., and Colett, T. S., 'The use of path integration to guide route learning in ants', *Nature*, 339 (1999): 6738.
Wehner, R., Harkness, Robert D., and Schmid-Hempel, P., 'Foraging strategies in individually searching ants', *Information Processing in Animals, Akademie der Wissenschaften und der Literatur*, 1 (1983): 1–79.
Wohlgemuth, S., Ronacher. R., and Wehner, R., 'Ant odometry in the third dimension', *Nature*, 411 (2001): 795–8.

On Atta

Mueller, U. G., Rehner, S. A., and Schultz, T. R., 'The evolution of agriculture in insects', *Science*, 281 (1998): 2034–8.
Mueller, U. G., Gerardo, N. M., Aanen, D. K., Six, D. L., and Schultz, T. R., 'The evolution of agriculture in insects', *Annual Review of Ecology, Evolution, and Systematics*, 36 (2005): 563–95.
Poulsen, M., and Boomsma, J. J., 'Mutualistic fungi control crop diversity in fungus-growing ants', *Science*, 307 (2005): 741–4.

On mutualism

Dejean, A., Solano, J. S., Ayroles, J., Corbara, B., and Orivel, J., 'Insect behaviour: arboreal ants build traps to capture prey', *Nature*, 434 (2005): 973.
Frederickson, M. E., Greene, M. J., and Gordon, D. M., '"Devil's garden" bedevilled by ants', *Nature*, 437 (2005): 495–6.

On pests

Holway, D. A., Lach, L., Suarez, A. V., Tsutsui, N. D., and Case, T. J., 'The causes and consequences of ant invasions', *Annual Review of Ecology and Systematics*, 33 (2002): 181–233.

Tschinkel, W., *The Fire Ant* (Cambridge, MA: Harvard University Press, 2006).

Wilson, E. O., 'Early ant plagues in the New World', *Nature*, 433 (2005): 32.

On kith and kin

Cahan, S. H., and Keller L., 'Complex hybrid origin of genetic caste determination in harvester ants', *Nature*, 424 (2003): 306–9.

Fournier, D., Estoup, A., Orlivel, J., Foucaud, J., Jourdan, H., Le Breton, J., and Keller L., 'Clonal reproduction by males and females in the little fire ant', *Nature*, 435 (2005): 1230–4.

Keller L., and Chapuisat, M., 'Cooperation among selfish individuals in insect societies', *BioScience*, 49 (1999): 899–909.

Langer, P., Hogendoorn, K., and Keller, L., 'Tug-of-war reproduction in a social bee', *Nature*, 428 (2004): 844–7.

Meunier, J., West, S. A., and Chapuisat, M., 'Split sex ratios in the social Hymenoptera: a meta-analysis', *Behavioural Ecology*, 19 (2008): 382–90.

Passera, L., Aron. S., Vargo, E. L., and Keller, L., 'Queen control of sex ratio in fire ants', *Science*, 293 (2001): 1308–10.

Pearcy, M., Aron, S., Doums, C., and Keller, L., 'Conditional use of sex and parthenogenesis for worker and queen production in ants', *Science*, 306 (2004): 1780–3.

Ratnieks, F. L. W., Foster, K. R., and Wenseleers, T., 'Conflict resolution in insect societies', *Annual Review of Entomology*, 51 (2006): 581–608.

Sundström, L., Chapuisat, M., and Keller, L., 'Conditional manipulation of sex ratios by ant workers: a test of kin selection theory', *Science*, 274 (1996): 993–5.

Trivers, R. L., and Hare, H., 'Haplodiploidy and the evolution of the social insects', *Science*, 191 (1976): 249–63.

On ants and IT

Bonabeau, E., Dorigo, M., and Théraulaz, G., *Swarm Intelligence: from Natural to Artificial Systems* (Oxford: Oxford University Press, 1999).

Bonabeau, E., and Théraulaz, G., 'Swarm smarts', *Scientific American* 282 (2000): 72–9.

Detrain, C., Deneubourg, J.-L., and Pasteels, J. M., *Information Processing in Social Insects* (Basel: Birkhäuser, 1999).

Krieger, M. J. B., Billeter, J.-B., and Keller, L., 'Ant-like task allocation and recruitment in cooperative robots', *Nature*, 404 (2000): 992–5.

Species Index

General Index